Lecture Notes in Engineering

The Springer-Verlag Lecture Notes provide rapid (approximately six months), refereed publication of topical items, longer than ordinary journal articles but shorter and less formal than most monographs and textbooks. They are published in an attractive yet economical format; authors or editors provide manuscripts typed to specifications, ready for photo-reproduction.

Lecture Notes in Engineering

Edited by C. A. Brebbia and S. A. Orszag

74

C. V. Camp, G. S. Gipson

Boundary Element Analysis of Nonhomogeneous Biharmonic Phenomena

Springer-Verlag Berlin Heidelberg GmbH

Series Editors
C. A. Brebbia · S. A. Orszag

Consulting Editors
J. Argyris · K.-J. Bathe · A. S. Cakmak · J. Connor · R. McCrory
C. S. Desai · K.-P. Holz · F. A. Leckie · G. Pinder · A. R. S. Pont
J. H. Seinfeld · P. Silvester · P. Spanos · W. Wunderlich · S. Yip

Authors
Dr. Charles V. Camp
Dept. of Civil Engineering
Memphis State University
Memphis/Tennessee 38152
USA

G. Steven Gipson
School of Civil Engineering
Oklahoma State University
Stillwater, Oklahoma 74074
USA

ISBN 978-3-540-55020-4   ISBN 978-3-642-84701-1 (eBook)
DOI 10.1007/978-3-642-84701-1

© Springer-Verlag Berlin Heidelberg 1992
Originally published by Springer-Verlag Berlin Heidelberg New York in 1992

Typesetting: Camera ready by author
61/3020-5 4 3 2 1 0   Printed on acid-free paper.

DEDICATION

From CVC:

To Kay

From GSG:

To the students and faculty of Byram High School in the late sixties, a magic environment where everyone could be a star. I would especially like to mention the following very special people:

two ladies:	Sue Robertson, Sherrie Perkins
three guys:	Gary Tipton, Roger Setzler, Norman Covington
five teachers:	Mrs. Emma Grace Cochran, Mrs. Velma Davis, Mrs. Frances Douglas, Mrs. Frances Dunkin, and Mrs. Marjorie Monk.

and "the principle of the thing": Mr. William H. Cochran.

Go Bulldogs!

PREFACE

At the date of this writing, there is no question that the boundary element method has emerged as one of the major revolutions on the engineering science of computational mechanics. The emergence of the technique from relative obscurity to a cutting edge engineering analysis tool in the short space of basically a ten to fifteen year time span is unparalleled since the advent of the finite element method. At the recent international conference BEM XI, well over one hundred papers were presented and many were published in three hard-bound volumes. The exponential increase in interest in the subject is comparable to that shown in the early days of finite elements. The diversity of applications of BEM, the broad base of interested parties, and the ever- increasing presence of the computer as an engineering tool are probably the reasons for the upsurge in popularity of BEM among researchers and industrial practitioners.

Only in the past few years has the BEM audience become large enough that we have seen the development of specialty books on specific applications of the boundary element method. The present text is one such book. In this work, we have attempted to present a self-contained treatment of the analysis of physical phenomena governed by equations containing biharmonic operators. The biharmonic operator defines a very important class of fourth-order PDE problems which includes deflections of beams and thin plates, and creeping flow of viscous fluids.

Before we detail the contents of the book, a few words on the state-of BEM text literature seem appropriate. It is fair to say that we are presently at a "thermal triple point" stage in BEM development. The boundary element research community consists of: (a) war-proven veterans of the technique who pioneered the practical use of boundary integral methods, mostly in the last five to fifteen years, but in some cases as far back as twenty-five years or more; (b) advanced practitioners who have learned the method for a specific advanced use; and (c) a vast number of enthusiastic well-educated persons who are striving to learn the basics of the method for the first time so as to expand their arsenal of numerical analysis weapons.

It is interesting to survey the textbook literature on boundary elements and not the variation in styles, philosophies, levels of sophistication, and attitudes about the method. Normally, the books are written with one of the three aforementioned audiences in mind. The tone taken in the present text is to attempt to address the needs of all three of the audiences, hopefully without over- or underwhelming any of the three. Since this is not strictly a teaching text, readers who have some familiarity with BEM and integral equation theory will feel most comfortable with the material. It is suggested that new arrivals to the BEM family look at the second author's earlier text *Boundary Element Fundamentals - Basic Concepts and Recent Developments in the Poisson Equation* (Topics in Engineering series, Vol. 2, Computational Mechanics Publications, 1987) for a simple treatment and an introduction to BEM jargon.

In the present text, we begin by introducing the nonhomogeneous biharmonic equation and its most famous one-dimensional example, the deflection of the long, straight elastic beam. This instructive example details the idea of a Green's function, shows how reciprocity may be used to derive the boundary integral equations for the beam, demonstrates the differences between finite elements and boundary elements for such an analysis, explains graphically why BEM assembly matrices are not symmetric, and, maybe most importantly, shows how finite elements (symmetric matrices and all) can be arrived at as a special case of BEM.

After this indoctrination, the text proceeds to the integral equation analysis of two-dimensional biharmonic phenomena and details the synthesis of the technique. The evolution from pure integral equation to the numerial technique of boundary elements is presented.

The next chapter discusses the implementation of the various element types available - constant, linear, quadratic, and an exciting new cubic spline, the "Overhauser element", which maintains local interelement derivative continuity without procedural complications. Explicit formulas of integrating these elements are derived.

The text then proceeds into the sensitive issue of domain integration for nonhomogeneous problems. The so-called problems associated with performing domain integrals in most governing equations other than those containing exclusively homogeneous elliptic PDE's is probably the major criticism of BEM in comparison with finite elements and finite differences. ("If one has to create a mesh anyway, why not use a mathematically less complex formulation?" the critic asks reasonably). We have attempted to use the power of the computer to the full for this purpose, and hence perform the domain integrals without explicit domain discretization. Several established techniques for avoiding this nuisance are discussed. One method, which has been used in some sense in other works, and which the authors have chosen to term "domain fanning" has been formalized as a general domain integration device. Explicit formulas for internal point calculations have been developed for each element type in cases where the problem geometry is rectilinear.

The following chapter then puts the theory to work by presenting examples from plate theory and creeping flow of viscous fluids. Several of the presented problems are nonlinear or may be treated as nonlinear for the purposes of calculation. These include plates on elastic foundations, plates of varying thickness, and those with inplane forces, as well as Stokes flow with non-zero Reynolds numbers. These problems are solved by iterating on the non-homogeneous solution technique, with the pseudo-nonlinear terms being represented as part of the nonhomogeneous term. BEM is shown to be a remarkably accurate device for solving this type of problem.

VII

The computer program used to solve the linear examples is presented in the next chapter. The program, written in VAX FORTRAN 77, is suitable for solving any two-dimensional biharmonic problem with arbitrary source terms. The structure of the program is described in detail with a preface to each subroutine, and a detailed set of user instructions. The reader who is not interested in typing in the program himself should order a copy of the program on diskette from the authors at the institution address for $50 American. The full nonlinear version of the program is available for $100.

The text concludes with a summary of the work and recommendations for further work in this area.

This monograph represents the combined efforts of both authors over several years of concentrated research in the general area of potential field problems in engineering. However, as anyone who has ever embarked on the synthesis of a research monograph well knows, credit is usually due to may people who are not listed as authors. The present work is hardly an exception to the rule. In no particular order, the authors wish to express their appreciation to a variety of people who directly of indirectly made this work possible. Many thanks are due to the faculty and staff of the Computer Graphics and Applications Laboratory at Louisiana State University. The fundamental computer work was developed under their auspices while the authors were in residence there. Special thanks are extended to CGRAL members Professor John A. Brewer III, Professor Andrew J. McPhate, Mr. Jeffrey N. Jortner, and Mr. Alaric Haag for their support and encouragement. Additionally, while thanking persons at LSU, the authors would like to mention Professor Danny D. Reible of the Chemical Engineering Department for his help and inspiration in attacking the creeping flow problem by boundary elements. Also to be thanked are Mr. Juan C. Ortiz and Mr. Harold Walters, who, besides pioneering the use of the Overhauser element in BEM, have provided insightful contributions to this work. Last, but not least in the way of technical help, we must acknowledge the efforts of Ms. Charlene Fries of the School of Civil Engineering at Oklahoma State University.

In the way of financial support for this work, the authors would like to express their gratitude to: Dr. N. Radhakrishnan and Mr. Paul Senter of the United States Corps of Engineers Waterways Experiment Station for their financial support; the Oklahoma State University Center for Engineering Research Grant #1-1-50728; and the OSU School of Civil Engineering for subsidizing the work with computer funds.

Charles V. Camp and G. Steven Gipson
Stillwater, Oklahoma, USA
December 1989

A Special Note from GSG:

A very special debt of gratitude is extended from the second author to the people for whom he has dedicated his portion of the book. Recently, I had occasion through some rather unique circumstances to find out exactly how profound was the influence of certain people I knew during my teen years; I had more than my fair share of good influences that complemented one another in a most miraculous way. To the special folks at good old

BHS for the legacies they left to me during our association in the late sixties - the students: Sue (for sound advice and inspiration), Sherrie (for inspiration and quality), Gary (for quality and perseverance), Roger (for perseverance and self-assurance), Norman (for self-assurance and for being a heck of a buddy through thick and thin); the teachers: Marjorie (a Latin teacher who taught me to be cultured and read), Velma (an English teacher who taught me to read and write), Frances (a history teacher who taught me to write and research), Frances (a librarian who taught me to research and think), Emma Grace (a math teacher who taught me to think and think), and William (the principal who held the ship together and on a steady course under some incredibly trying circumstances) - an undying thanks from the "monster" you helped to create.

Table of Contents

CHAPTER 1

BOUNDARY ELEMENTS AND THE BIHARMONIC EQUATION

1.1   Introduction

Possibly the most exciting development in applied numerical methods over the last fifteen years has been the popularization of "boundary element" methods. This modern engineering analysis technique has evolved from the much older mathematical subjects of integral equations, Green's' functions, and potential theory. The most graphic success of boundary elements have been in the analyses of scalar potential problems, most notably phenomena governed by Laplace's equation ($\nabla^2 \psi = 0$) or the more general Poisson equation ($\nabla^2 \psi = f$). Other successful applications have been realized primarily in elliptic boundary value problems such as elastostatics analysis, steady-state wave propagation, etc.

Relatively little attention has been devoted to the direct boundary integral analysis of phenomena governed by biharmonic operators ($\nabla^4 = \nabla^2 \nabla^2$). There are a variety of practical physical problems that fall under this heading. Examples include elastic beam deflections due to bending, deflections of thin elastic plates, and creeping flow of viscous fluids. All these examples have a number of variants which may be classified as being linear or nonlinear, homogeneous or nonhomogeneous, or combinations thereof depending upon the degree of refinement that the analyst desires. The purpose of this text is to present a detailed prescription for solving steady-state equations of the form:

$$\nabla^4 \psi = f \qquad\qquad (1.1)$$

in which the function f is an arbitrary function of the spatial coordinates. Equation (1.1) will be known generically as the "non-homogeneous biharmonic equation" throughout this text. It will be shown that the more general case, in which f is a function of the field variables and its various partial derivatives, may be solved by iterating on the procedure used to solve Equation (1.1).

This text proceeds by presenting the necessary theory for converting the partial differential Equation (1.1) into an integral equation. The various fundamental solutions are derived and the transition from the analytical formulation to the numerical procedure is developed. Much discussion is given to the appropriate element types for biharmonic analyses. Particular attention is given to a remarkable new cubic splining element known as the Overhauser element. Also, a chapter is devoted to the issue of domain integration. This is a troublesome point in boundary elements because its tactical advantage over finite elements is severely reduced when domain integrals must be performed. Several alternate procedures that regain a great deal of that advantage will be detailed. Example biharmonic analyses from the linear and nonlinear theories of thin plate bending, as well as several applications to creeping flow of viscous fluids are presented. A computer program, listed in VAX FORTRAN 77 and designed to solve the linear nonhomogeneous form of Equation (1.1), is described and documented in the following chapter. Finally, a summary and conclusions section is presented.

We begin the introduction to the analysis of the biharmonic equation with boundary elements by considering perhaps the simplest example -- deflections of linear elastic beams. This example is most instructive since it shows the detailed derivation and meaning of the Green's function, demonstrates poignant differences between finite elements and boundary elements as far as philosophy, and illustrates the reasons behind the often controversial fact that the BEM matrices are not symmetric. Also, the reduction of the boundary element method to the finite element method points out the latter technique as a special case of BEM.

3

1.2 The Elastic Beam

1.2.1 Introduction

Consider small static displacements of a long, straight, homogeneous elastic beam of length L, and flexural rigidity EI. Let x denote the coordinate along the length of the beam, with the origin at the left end, and define a positive deflection as being upwards. The beam carries an arbitrary transverse loading $p(x)$, which gives rise to reactive shearing forces and bending moments at the ends. Using the standard civil engineering convention for the definition of positive quantities the general situation appears as shown in Figure 1.1. It should be noted that this beam could possibly be an excised segment of a longer beam, a fact which is important in the derivations to come. In that case, the reactive shears V_1 and V_2, and reactive bending moments M_1 and M_2, shown in Figure 1.1 would be simply internal reactions. Compatibility and equilibrium of the larger elastic system would have to be maintained at $x = 0$ and $x = L$.

The popular "finite element" method, in principle, works in the manner just described; small pieces of the overall body are patched together at nodal points, and certain rules of compatibility are enforced at those points. For a beam finite element, continuity of the deflection and its first derivative (slope) are

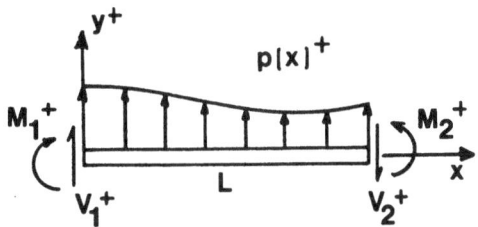

Figure 1.1   Nomenclature and Definitions of Positive
Quantities for Beam Analysis.

mandated. By defining suitable splining shape functions, the following matrix relationship, using the notation of Figure 1.2, may be derived for a beam finite element (i.e., Weaver and Johnson, 1984)

$$\frac{EI}{L^3} \begin{bmatrix} 12 & 6L & -12 & 6L \\ 6L & 4L^2 & -6L & 2L^2 \\ -12 & -6L & 12 & -6L \\ 6L & 2L^2 & -6L & 4L^2 \end{bmatrix} \begin{Bmatrix} y_i \\ \theta_i \\ y_j \\ \theta_j \end{Bmatrix} = \begin{Bmatrix} P_i \\ M_i \\ P_j \\ M_j \end{Bmatrix} \qquad (1.2)$$

(Note the difference in the notation between Figures 1.1 and 1.2)

Relationship (1.2) is well-known to every student of structural mechanics. In finite element parlance, the 4 x 4 matrix is the "stiffness" matrix for the system. It relates the nodal displacements and forces, where the terms "displacements" and "forces" are to be interpreted in a generalized sense. The stiffness matrix for a conservative system can always be written as a symmetric matrix. The symmetry follows from basic principles of work and energy, and is one of the most useful and exploited properties of finite element matrices.

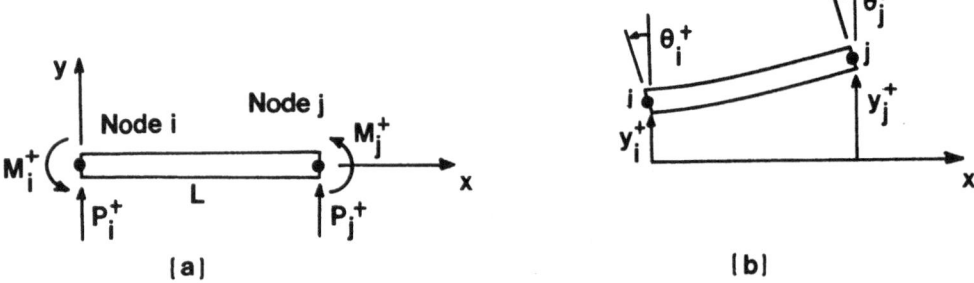

Figure 1.2 (a) A Finite Element for a Beam Showing Typical Sign Conventions for the End Forces and Couples: Any Transverse Loading is Absorbed into the Nodal Forces; (b) The Four Degrees of Freedom for Each Element Consist of the Two Nodal Deflections and the Two End Rotations.

In this section, the application of boundary elements to the beam problem is discussed. Boundary element formulations do not always lead to symmetric matrices, even in elastic systems. The reasons for this are several and perfectly justified, but the newcomer to BEM often finds this fact troublesome. The usual culprit leading to nonsymmetry in elasticity by BEM is the basic set of unknown quantities. For instance, in two and three dimensional elasticity, application of the boundary element procedure results in a solution for nodal displacements and <u>surface tractions</u>. This should be contrasted to finite elements, in which the solution set consists of unknown nodal displacements and <u>forces</u>. That is, in the two formulations, one set of basic unknown quantities, the tractions in BEM and the forces in FEM, are not physical equivalents. The FEM scenario, as stated earlier, can be cast in terms of a symmetric stiffness matrix. It is important to understand the distinction of the word "can" in the previous sentence. The stiffness matrix can also <u>not</u> be written as a symmetric relationship by simply redefining the basic parameters. For instance, redefining θ_i as being positive in a clockwise sense in Figure 1.2(b) without changing the sense of the couple M_i in Figure 1.2(a) negates the symmetry of Equation (1.2). An easier way to achieve nonsymmetry is to simply interchange two rows in Equation (1.2). The convenience of symmetric matrices is so acute that no analyst ever considers writing the basic equations any other way. However, symmetry of a matrix must usually be coaxed in some fashion; it is rarely a given state of affairs.

On the other hand, if the system is conservative, it is always possible to write the boundary element equations as symmetric matrices by a suitable redefinition of the basic variables. However, it is not always convenient or straightforward, and never necessary, to do so. The beam problem, however is one formulation where the fundamental set of unknown quantities in FEM and BEM are physically equivalent; still, the formulations for BEM and FEM are not mathematically identical. In the next section, it will be shown that the boundary element formulation gives the equivalent FEM stiffness matrix.

1.2.2 Boundary Element Analysis

It is assumed that the reader is familiar with the derivation of the finite element beam stiffness matrix. The following formulation should be compared with finite elements in terms of philosophy, mathematical sophistication, etc., as the reader peruses the derivation.

The boundary element method begins by considering the domain of the problem in its entirety without subdivisions. All the discretization is performed on the boundary. In this case, there are but two "nodes" involved, regardless of the complexity of the problem. The two nodes are placed, respectively, at x=0 and at x=L in Figure 1.1.

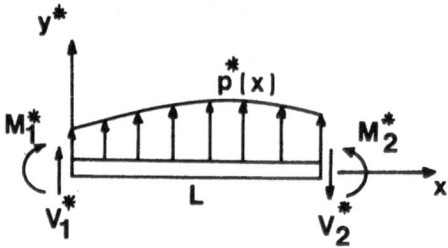

Figure 1.3   An Imaginary Loading System on the Beam of Figure 1.1

The boundary element relationship for the beam may be approached by several different avenues. Perhaps the simplest is to invoke the Somigliana identity, which is the continuous form of Betti's reciprocity theorem (Banerjee and Butterfield, 1981). Referring to Figure 1.3, consider a hypothetical loading system on the beam of Figure 1.1, denoted by starred (*) quantities. The Somigliana identity is:

$$\int_0^L p(x) \, y^* \, dx - M_1 y_i^{*\prime} + V_1 y_1^* + M_2 y_2^{*\prime} - V_2 y_2^*$$

$$= \int_0^L p^*(x) \, y \, dx - M_1^* y_1^{\prime} + V_1^* y_1 + M_2^* y_2^{\prime} - V_2^* y_2 \tag{1.3}$$

where the primes denote differentiation with respect to x. In other words, Equation (1.3) states that the sum of the real forces (in a generalized context) times the virtual displacements equals the sum of the virtual forces times the real displacements. Equation (1.3) is fundamental in structural mechanics. The equation has the potential of leading to a variety of different common analysis techniques. The boundary element formulation follows from dictating that the virtual distributed load be a single concentrated force of unit intensity acting at an arbitrary point x_ℓ on the beam. Mathematically, this means that the loading should be represented by the Dirac delta function (Figure 1.4). That is,

$$p^*(x) = \delta(x-x_\ell) \tag{1.4}$$

Substituting this into Equation (1.3) and using the selective property of the Dirac delta, results in

$$y(x_1) = M_1^* y_1' - V_1^* y_1 - M_2^* y_2' + V_2^* y_2 + \int_0^L p(x) y^* dx$$

$$+ M_1 y_1^{*'} + V_1 y_1^* + M_2 y_2^{*'} - V_2 y_2^* \tag{1.5}$$

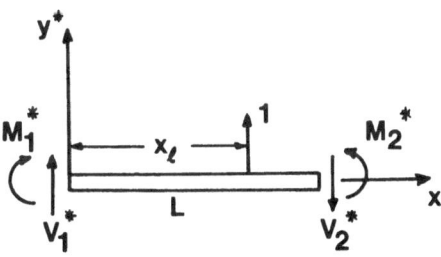

Figure 1.4   The Definition of $p^*(x)$ Leading
to Boundary Elements

which is a solution for the deflection at an arbitrary point x_ℓ once the boundary values of y and y' are determined. The remaining starred quantities in Equation (1.5) are now defined by the loading function (1.4). To find the explicit values, the solution of the ordinary differential equation governing the deflection of beams in the starred system should be determined:

$$EIy^{*iv} = \delta(x-x_\ell) \tag{1.6}$$

One integration of Equation (1.6) gives:

$$EIy^{*'''} = H(x-x_\ell) + C_1 = V^* \tag{1.7}$$

where $H(x-x_\ell)$ denotes the Heaviside step function. [For the reader who is not familiar with mathematical manipulations of these generalized functions, a review of a text on applied mathematics or one containing a discussion of fundamental solutions may be worthwhile (i.e., Cushing, 1975 or Gipson, 1987).]

The boundary conditions on Equation (1.7), which physically represents the relationship between the shearing force and the change in beam curvature, are fairly obscure. This is where visualizing the beam to be a piece of a much larger beam, in fact, an infinitely long beam, may be of help. The reactive shear exercises a jump discontinuity at a point where an applied load exists. The shearing force is therefore antisymmetric in x and x_ℓ. That is:

$$EIy^{*'''}(x,x_\ell) = - EIy^{*'''}(x_\ell,x) \tag{1.8}$$

This is quite easy to see if one draws a shear diagram for an infinitely long beam loaded by a single concentrated force at $x = x_\ell$. For the sake of easy visualization in this exercise, one should pretend that the beam is simply supported at $\pm\infty$, although it doesn't really matter how the beam is supported. Applying the result of Equation (1.8) to Equation (1.7) gives

$$H(x-x_\ell) + C = -H(x_\ell-x) - C_1$$

$$C_1 = \frac{-H(x-x_\ell) - H(x_\ell-x)}{2} \tag{1.9}$$

Notice that in this format of liberally using generalized functions, C_1 actually represents either of two constants; C_1 takes on either value according to the location of x_ℓ. Therefore:

$$v^* = EIy^{*'''} = H(x-x_\ell) - \frac{H(x-x_\ell) + H(x_\ell-x)}{2}$$

$$= \frac{H(x-x_\ell) - H(x_\ell-x)}{2} = \frac{1}{2}\,sgn(x-x_\ell) \tag{1.10}$$

where "sgn" denotes the "sign" function, and is interpreted as being equal to the sign of its argument times unity. Another integration of Equation (1.8) gives the bending moment.

$$EIy^{*''} = \frac{(x-x_\ell)\,H(x-x_\ell) + (x_\ell-x)\,H(x_\ell-x)}{2} + C_2$$

$$= \frac{|x-x_\ell|}{2} = C_2 = M^* \tag{1.11}$$

Reverting to our infinite beam caricature, it is concluded that the bending moment in the beam must be symmetric in x and x_ℓ. Thus:

$$EIy^{*''}(x,x_\ell) = EIy^{*''}(x_\ell,x)$$

$$\frac{|x-x_\ell|}{2} + C_2 = \frac{|x_\ell-x|}{2} + C_2$$

which leaves the value of C_2 inconclusive. Apparently, C_2 has the role of taking on the value of some fixed end moment at $\pm\infty$. The possibility of its existence throughout the next integration sequence will be maintained. Therefore

$$EIy^{*'} = \frac{(x-x_\ell)^2}{4}\,sgn(x-x_\ell) + C_2x + C_3 \tag{1.12}$$

which is proportional to the slope of the beam. Again, physical considerations mandate that the slope of the beam caused by the unit load must be antisymmetric in x and x_ℓ. Therefore:

$$y^{*'}(x,x_\ell) = -\,y^{*'}(x_\ell,x)$$

$$\frac{(x-x_\ell)^2}{4}\,sgn(x-x_\ell) + C_2x + C_3 = -\,\frac{(x_\ell-x)^2}{4}\,sgn(x_\ell-x) - C_2x_\ell - C_3$$

which is possible if and only if $C_2 = C_3 = 0$.

The deflection of the beam is the result of the next integration:

$$EIy^* = \frac{|x-x_\ell|^3}{12} + C_4 \tag{1.13}$$

The constant C_4 obviously represents a rigid body translation of the beam. Since the reference configuration for the beam is arbitrary, C_4 is simply set to zero. The function

$$y^* = \frac{|x-x_\ell|^3}{12EI} \tag{1.14}$$

is known as the "fundamental solution" for the beam, since any deflection of the beam due to any transverse loading may be represented as a superposition of deflections like those given by Equation (1.14). Equation (1.14) is also called the Green's function for the infinite beam due to a unit loading.

With Equations (1.10), (1.11), (1.12), and (1.14), the boundary values of the starred quantities in Equation (1.15) may be computed. Explicitly, for $0 < x_\ell < L$, we find

$$V_1^* = EIy^{*'''}(0,x_\ell) = \frac{\text{sgn}(0-x_\ell)}{2} = \frac{1}{2} \tag{1.15}$$

$$M_1^* = EIy^{*''}(0,x_\ell) = \frac{|0-x_\ell|}{2} = \frac{x_\ell}{2} \tag{1.16}$$

$$V_2^* = EIy^{*'''}(L,x_\ell) = \frac{\text{sgn}(L-x_\ell)}{2} = \frac{1}{2} \tag{1.17}$$

$$M_2^* = EIy^{*''}(L,x_\ell) = \frac{|L-x_\ell|}{2} = \frac{L-x_\ell}{2} \tag{1.18}$$

$$y_1^{*'} = y^{*'}(0,x_\ell) = \frac{(0-x_\ell)^2}{4EI}\,\text{sgn}(0-x_\ell) = \frac{-x_\ell^2}{4EI} \tag{1.19}$$

$$y_\ell^* = y^*(0,x_\ell) = \frac{|0-x_\ell|^3}{12EI} = \frac{x_\ell^3}{12EI} \tag{1.20}$$

$$y_2^{*'} = y^{*'}(0,x_\ell) = \frac{(L-x_\ell)^2}{4EI}\,\text{sgn}(L-x_\ell) = \frac{(L-x_\ell)^2}{4EI} \tag{1.21}$$

$$y_2^* = y^*(0,x_\ell) = \frac{|L-x_\ell|^3}{12EI} = \frac{(L-x_\ell)^3}{12EI} \tag{1.22}$$

Thus, the "boundary integral" relationship, Equation (1.5), becomes a function of the coordinate 'x_ℓ':

$$y(x_\ell) = \frac{x_\ell y_1'}{2} + \frac{y_1}{2} + \frac{(x_\ell-L)}{2} y_2' + \frac{y_2}{2} + \frac{M_1 x_\ell^2}{4EI} + \frac{V_1 x_\ell^3}{12EI}$$

$$= \frac{M_2(L-x_\ell)^2}{4EI} + \frac{V_2(x_\ell-L)^3}{12EI} + \int_0^L p(x) \frac{|x-x_\ell|^3}{12EI} dx \tag{1.23}$$

Actually, calling this a "boundary integral relationship" is engaging in a mild perversion of terms since there are no boundary integrals to evaluate. In one dimensional problems, the integrations on the boundary reduce to simple functional evaluations at the ends of the beam.

The boundary element model consists of the entire beam demarked by nodes at its two ends. In order to have a well-posed problem, it is necessary to have a certain number of boundary conditions specified. The boundary conditions are of two varieties. Specified values of the displacements (y) and slope (y') are known as "forced" boundary conditions. Specified values of the shearing forces (V) and the bending moments (M) are known as "natural" boundary conditions. A well-posed problem specifies either a shearing force or a displacement at a boundary point; and either a bending moment or a slope at the point. It is important to understand that both a shearing force and a displacement cannot be prescribed at the same point; neither can both a bending moment and a slope be specified. However, one or the other of each of these boundary conditions types must be specified for a unique solution to exist. In the ensuing analysis, the problem is assumed to be well-posed. Note that there are always four known boundary conditions and four unknown boundary conditions in a well-posed beam problem.

The boundary element method proceeds from this point by first solving for the unknown boundary data (that which has not been specified) in terms of that

which has been specified. To successfully implement this procedure, four simultaneous equations in the four unknown quantities must be obtained. Two of these
equations follow immediately from Equation (1.23). First, set $x_\ell = 0$ in Equation
(1.23). This gives $y(x_\ell = 0) = y_1$ as the left-hand side. This simplified equation is:

$$y_1 = -Ly_2' + y_2 + \frac{M_2 L^2}{2EI} - \frac{V_2 L^3}{6EI} + \int_0^L \frac{x^3 p(x)}{6EI} \, dx \qquad (1.24)$$

The other relation is obtained by setting $x_\ell = L$, giving $y(x_\ell = 0) = y_2$.

$$y_2 = Ly_1' + y_1 + \frac{M_1 L^2}{2EI} - \frac{V_1 L^3}{6EI} + \int_0^L \frac{(1-x)^3 p(x)}{6EI} \, dx \qquad (1.25)$$

The other two equations needed come after differentiating Equation (1.23) with
respect to x_ℓ The result is:

$$y'(x_\ell) = \frac{y_1'}{2} + \frac{y_2'}{2} + \frac{M_1 x_\ell}{2EI} + \frac{V_1 x_\ell^2}{4EI} - \frac{M_2(L-x_\ell)}{2EI}$$

$$+ \frac{V_2(x_\ell - L)^2}{4EI} + \int_0^L \frac{(x-x_\ell)^2 \, \mathrm{sgn}(x_\ell - x) \, p(x)}{4EI} \, dx \qquad (1.26)$$

Setting $x_\ell = 0$ in Equations (1.26) gives

$$y_1' = y_2' - \frac{M_2 L}{EI} + \frac{V_2 L^2}{2EI} - \int_0^L \frac{(x-L)^2 p(x)}{2EI} \, dx \qquad (1.27)$$

Setting $x_\ell = L$ gives:

$$y_2' = y_1' + \frac{M_1 L}{EI} + \frac{V_1 L^2}{2EI} + \int_0^L \frac{(x-L)^2 \, p(x)}{2EI} \, dx \qquad (1.28)$$

The system of equations given by Equations (1.24), (1.25), (1.27), and (1.28) may
be cast in the format

$$[H]\{Y\} = [G]\{F\} + \{B\} \qquad (1.29)$$

where the terms are defined as follows:

$$\{Y\} = \begin{Bmatrix} y_1 \\ y_1' \\ y_2' \\ y_2' \end{Bmatrix} \qquad \{F\} = \begin{Bmatrix} V_1 \\ M_1 \\ V_2 \\ M_2 \end{Bmatrix} \qquad (1.30)$$

$$[H] = \begin{bmatrix} 1 & 0 & -1 & L \\ 0 & 1 & 0 & -1 \\ -1 & -L & 1 & 0 \\ 0 & -1 & 0 & 0 \end{bmatrix} \qquad (1.31)$$

$$[G] = \frac{L}{6EI} \begin{bmatrix} 0 & 0 & L^2 & 3L \\ 0 & 0 & 3L & -6 \\ L^2 & 3L & 0 & 0 \\ 3L & 6 & 0 & 0 \end{bmatrix} \qquad (1.32)$$

and

$$\{B\} = \int_0^L \frac{p(x)}{6EI} \begin{Bmatrix} x^3 \\ -3x^2 \\ (1-x)^3 \\ 3(1-x)^2 \end{Bmatrix} dx \qquad (1.33)$$

Equation (1.29) is the usual form for boundary element matrices. Since unknown values of $\{Y\}$ align with known values of $\{F\}$ in Equation (1.29) and vice-versa, the system of equations may be reordered into the form:

$$[K]\{X\} = \{Z\} \qquad (1.34)$$

where $\{X\}$ contains all the unknown boundary data, and $\{Z\}$ contains the algebraic sum of products of all known boundary values with their corresponding columns of $[G]$ or $[H]$ as the case may be. It should be noted that neither $[H]$ nor $[G]$ are symmetric, and only by the most miraculous of miracles will $[K]$ be symmetric.

However, it should also be noted that none of these matrices is a "stiffness" matrix in the usual structures sense.

Once the system (1.34) is solved, all the boundary values will be known. These are substituted in to Equation (1.23) and the resulting equation can be evaluated at any point x_ℓ. The same is true for Equation (1.26). Also, Equation (1.26) may be differentiated further to obtain the shearing force and the bending moment at any point x_ℓ. It is here that the power of the boundary element method really becomes apparent. The solution is not limited to finding accurate values at nodal points and interpolating according to the dictates of the shape functions. Nothing has been approximated in the domain. Calculating the deflections, the slope, the bending moment, and the shear force at any point may be accomplished with equal precision. The concepts developed in this section are best illustrated by a practical example wherein the details associated with the procedure are presented.

1.2.3 Example - A Cantilevered Beam

The reader familiar with finite element analysis of beams should study, compare, and contrast this section with the equivalent FEM procedure. Consider a cantilevered beam of length L, and flexural rigidity EI carrying its weight described by w_0 in force/length, and a concentrated upward load at its end as shown in Figure 1.5. The specified boundary conditions are

$$y(0) \quad = y_1 = 0$$
$$y'(0) = y_1 = 0$$
$$M(L) \quad = M_2 = 0 \tag{1.35}$$
$$V(L) \quad = V_2 = -P_0$$

The unknown values $y(L) = y_2$, $y'(L) = y_2'$, $M(0) = M_1$, and $V(0) = V_1$ must be computed. Although when using manual methods, the algorithm described in the preceding section may not be the most efficient, it will nonetheless be used here for the purposes of illustration.

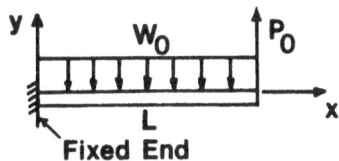

Figure 1.5 A Cantilever Under Concentrated
Load and Self-Weight

The first step is to set up the system of equations described by Equation (1.29). The [H] and [G] matrices look precisely the same as those given in Equations (1.31.) and (1.32). The force and displacement vectors, Equation (1.30) are upon the substitution of Equation (1.35):

$$\{Y\} = \begin{Bmatrix} 0 \\ 0 \\ y_2 \\ y_2' \end{Bmatrix} \qquad \{F\} = \begin{Bmatrix} V_1 \\ M_1 \\ -P_0 \\ 0 \end{Bmatrix} \tag{1.36}$$

The values of the vector defined in Equation (1.33) must be evaluated explicitly. The loading $p(x)$ is given by

$$p(x) = -w_0 \tag{1.37}$$

The concentrated load at the end of the beam has already been accounted for as a boundary value.

$$\{B\} = \int_0^L \frac{-w_0}{6EI} \begin{Bmatrix} x^3 \\ -3x^2 \\ (L-x)^3 \\ 3(L-x)^2 \end{Bmatrix} dx = \frac{-w_0}{6\ EI} \begin{Bmatrix} L^4/24 \\ -L^3 \\ L^4/24 \\ L^3 \end{Bmatrix} \tag{1.38}$$

The system of equations in raw form appears as:

$$\begin{bmatrix} 1 & 0 & -1 & L \\ 1 & 0 & -1 & L \\ -1 & -L & 1 & 0 \\ 0 & -1 & 0 & 1 \end{bmatrix} \begin{Bmatrix} 0 \\ 0 \\ y_2 \\ y_2' \end{Bmatrix} = \frac{L}{6EI} \begin{bmatrix} 0 & 0 & -L^2 & 3L \\ 0 & 0 & 3L & -6 \\ L^2 & 3L & 0 & 0 \\ 3L & 6 & 0 & 0 \end{bmatrix} \begin{Bmatrix} V_1 \\ M_1 \\ -P_0 \\ 0 \end{Bmatrix}$$

$$+ \frac{w_0 L^3}{6EI} \begin{Bmatrix} \frac{-L}{24} \\ 1 \\ \frac{-L}{24} \\ -1 \end{Bmatrix} \tag{1.39}$$

Equation (1.39) may be rearranged by performing the matrix multiplications involving known quantities and grouping the unknown quantities on the left-hand side. The result is:

$$\begin{bmatrix} 0 & 0 & -1 & L \\ 0 & 0 & 0 & -1 \\ \frac{-L^3}{2EI} & \frac{-L^2}{EI} & 1 & 0 \\ \frac{-L^2}{2EI} & \frac{-L}{EI} & 0 & 1 \end{bmatrix} \begin{Bmatrix} V_1 \\ M_1 \\ y_2 \\ y_2' \end{Bmatrix} = \begin{Bmatrix} \frac{P_0 L^2}{6EI} - \frac{w_0 L^4}{24EI} \\ \frac{-P_0 L^2}{2EI} + \frac{w_0 L^3}{6EI} \\ \frac{-w_0 L^4}{24EI} \\ \frac{-w_0 L^3}{6EI} \end{Bmatrix} \tag{1.40}$$

This system may be solved easily to yield

$$\begin{Bmatrix} V_1 \\ M_1 \\ y_2 \\ y_2' \end{Bmatrix} = \begin{Bmatrix} -P_0 + w_0 L \\ P_0 L - \frac{w_0 L^2}{2} \\ \frac{P_0 L^2}{3EI} - \frac{w_0 L^2}{8EI} \\ \frac{P_0 L^2}{2EI} - \frac{w_0 L^3}{6EI} \end{Bmatrix} \tag{1.41}$$

The validity of the results of Equation (1.41) may be easily verified by elementary analysis.

The deflections at any point along the beam may be calculated by reusing Equation (1.23) with the boundary values substituted. Equation (1.23) now appears as

$$y(x_\ell) = \frac{(x_\ell - L)}{2} \left(\frac{P_0 L^2}{2EI} - \frac{w_0 L^3}{6EI} \right) + \frac{1}{2} \left(\frac{P_0 L^3}{3EI} - \frac{w_0 L^4}{8EI} \right)$$

$$+ \left(P_0 L - \frac{w_0 L^2}{2} \right) \left(\frac{x_\ell^2}{4EI} \right) - (P_0 - w_0 L)\left(\frac{x_\ell^3}{12EI} \right)$$

$$- \frac{P_0 (x_\ell - L)^3}{12EI} - \int_0^L w_0 \frac{|x - x_\ell|^3}{12EI} \, dx \qquad (1.42)$$

For instance, to calculate the deflection at $x_\ell = L/2$, simply make the indicated substitution to get

$$y(L/2) = \frac{-L}{2} \left(\frac{P_0 L^2}{2EI} - \frac{w_0 L^3}{6EI} \right) + \frac{1}{2} \left(\frac{P_0 L^3}{3EI} - \frac{w_0 L^4}{8EI} \right)$$

$$+ \left(P_0 L - \frac{w_0 L^2}{2} \right) \left(\frac{L^2}{16EI} \right) - (P_0 - w_0 L)\left(\frac{L^3}{96EI} \right)$$

$$+ \frac{P_0 L^3}{96EI} - \int_0^L w_0 \frac{|x - \frac{L}{2}|^3}{12EI} \, dx$$

$$= \frac{5 P_0 L^3}{48EI} - \frac{w_0 L^4}{24EI} - \frac{w_0}{12EI} \frac{(x - \frac{L}{2})^4}{4} \, \text{sgn}(x - \frac{L}{2}) \,\Big|_0^L$$

$$= \frac{5 P_0 L^3}{48EI} - \frac{17 w_0 L^4}{384EI} \qquad (1.43)$$

another result easily verified by elementary methods. Equations (1.26) could just as easily be used to compute the slope. To get an explicit formula for the bending moment, simply differentiate Equation (1.26) once more and multiply by EI. The result is :

$$M(x_\ell) = \frac{M_1 + M_2}{2} + \frac{V_1 x_\ell}{2} + \frac{V_2 (x_\ell - L)}{2} + \int_0^L \frac{|x - x_\ell|}{2} p(x) \, dx \qquad (1.44)$$

At $x_\ell = L/2$,

$$M(L/2) = \frac{1}{2} \left(P_0 L - \frac{w_0 L^2}{2} \right) - \frac{L}{4} (P_0 - w_0 L) - \frac{w_0}{2} \int_0^L \left| x - \frac{L}{2} \right| dx$$

$$M(L/2) = \frac{P_0 L}{4} - \frac{w_0}{2} \frac{(x-L/2)^2}{2} \, sgn(x-1/2) \, \Big|_0^L$$

$$= \frac{P_0 L}{4} - \frac{w_0 L^2}{8} \, ,$$

another perfect result.

The boundary element method is philosophically different from the equivalent finite element displacement method. Mechanically speaking, BEM requires more mathematical finesse and more outside preprocessing before a general program can be written. However, this is compensated for in part by the excellent quality of the results due to the exact representation of the loading and the lack of approximation of the solution by shape functions. Also worthy of note is the fact that there is absolutely no concern about the grade of the mesh in the BEM results.

The finite element method is, however, a limiting case of boundary elements; and therefore FEM should be derivable from BEM. The equivalence of the two techniques in the beam analysis is the subject of the next section.

1.2.4 A FEM Stiffness Matrix from BEM

As mentioned previously, the basic set of unknown quantities to be solved for in BEM need not correspond to the generalized forces and displacements that are the rule in finite elements. However, this is not the case for the beam. The boundary variables are transverse displacements (y), angular displacements (y'), shearing forces (V), and the bending moment (M). Although they differ in sign and in the manner in which they are composed, the forces correspond to forces, angles to angles, etc. Therefore, it should be relatively simple to prove that the BEM formulation gives rise to a symmetric stiffness matrix equivalent in every way to that derived in finite elements.

Consider Equation (1.29) in which a BEM matrix relationship between the generalized displacements {y} and the generalized forces {F} are given:

$$[H]\{Y\} = [G]\{F\} + \{B\} \qquad (1.29)$$

The fact that Equation (1.29) is a global relationship in no way limits the validity of the ensuing analysis. A free body diagram of the entire beam or any section of the beam appears just as shown in Figure 1.1. Therefore, a segment of the beam that happens to coincide with a finite element representation of the same segment will be considered.

The required stiffness relationship apparently follows from simply multiplying Equation (1.29) through by $[G]^{-1}$. Therefore:

$$([G]^{-1}[H]) \ \{Y\} = \{F\} + [G]^{-1}\{B\} \qquad (1.45)$$

The term relating to the stiffness matrix may be immediately identified as:

$$[A] \equiv [G]^{-1}[H] \qquad (1.46)$$

whereas the load vector is the right-hand side of Equation (1.45):

$$\{P\} = \{F\} + [G]^{-1}\{B\} \qquad (1.47)$$

Next, it will be shown that the [A] matrix and the P vector are those explicitly defined by Equation (1.2). To do this, the [G] matrix must be explicitly inverted. This is fairly easy to do with a matrix such as [G], Equation (1.32), which may be written in partitioned form as:

$$[G] = \begin{bmatrix} [0] & [G_1] \\ [G_2] & [0] \end{bmatrix} \qquad (1.48)$$

The inverse of $[G]$ is obviously

$$[G]^{-1} = \begin{bmatrix} [0] & [G_2]^{-1} \\ [G_1]^{-1} & [0] \end{bmatrix} \qquad (1.49)$$

Therefore, the problem of inversion reduces to the simple problem of inverting two 2 x 2 matrices. Noting that

$$[G_1] = \frac{L}{6EI} \begin{bmatrix} -L^2 & 3L \\ 3L & -6 \end{bmatrix} \qquad (1.50)$$

it is easy to prove that

$$[G_1]^{-1} = \frac{2EI}{L^3} \begin{bmatrix} 6 & 3L \\ 3L & 6 \end{bmatrix} \qquad (1.51)$$

Likewise,

$$[G_2] = \frac{L}{6EI} \begin{bmatrix} L^2 & 3L \\ 3L & 6 \end{bmatrix} \qquad (1.52)$$

implying that

$$[G_2]^{-1} = \frac{2EI}{L^3} \begin{bmatrix} -6 & 3L \\ 3L & -L^2 \end{bmatrix} \qquad (1.53)$$

Now, note that $[H]$, Equation (1.31) may be written as

$$[H] = \begin{bmatrix} [I] & [H_1] \\ [H_2] & [I] \end{bmatrix} \qquad (1.54)$$

Therefore,

$$[A] = [G]^{-1} [H] = \begin{bmatrix} [0] & [G_2]^{-1} \\ [G_1]^{-1} & [0] \end{bmatrix} \begin{bmatrix} [I] & [H_1] \\ [H_2] & [I] \end{bmatrix}$$

$$= \begin{bmatrix} [G_2]^{-1} [H_2] & [G_2]^{-1} \\ [G_1]^{-1} & [G_1]^{-1} [H_1] \end{bmatrix} \qquad (1.55)$$

Forming the various matrix products from Equations (1.51) and (1.53), and identifying the terms of Equations (1.54) from Equation (1.31), results in:

$$[G_2]^{-1} [H_2] = \frac{2EI}{L^3} \begin{bmatrix} -6 & 3L \\ 3L & -L^2 \end{bmatrix} \begin{bmatrix} -1 & -L \\ 0 & -1 \end{bmatrix} = \frac{2EI}{L^3} \begin{bmatrix} 6 & 3L \\ -3L & -2L^2 \end{bmatrix} \qquad (1.56)$$

$$[G_1]^{-1} [H_1] = \frac{2EI}{L^3} \begin{bmatrix} 6 & 3L \\ 3L & L^2 \end{bmatrix} \begin{bmatrix} -1 & L \\ 0 & -1 \end{bmatrix} = \frac{2EI}{L^3} \begin{bmatrix} -6 & 3L \\ -3L & 2L^2 \end{bmatrix} \qquad (1.57)$$

By placing Equation (1.56) and (1.57) into Equation (1.55) along with Equations (1.51) and (1.53), the final form of [A] is obtained:

$$[A] = \frac{2EI}{L^3} \begin{bmatrix} 6 & 3L & -6 & 3L \\ -3L & -2L^2 & 3L & -L^2 \\ 6 & 3L & -6 & 3L \\ 3L & L^2 & -3L & 2L^2 \end{bmatrix} \qquad (1.58)$$

Equation (1.58) is not quite identical to the stiffness matrix of Equation (1.2) because the senses on the generalized displacements and forces are not identical in the {Y} and {F} vectors. By referring to Figures 1.1 and 1.2, it may be seen that y_1' and θ_i differ by a sign. The term y_1' is the actual slope of the curve defining the beam in the x,y frame, whereas, θ_i is the angle of rotation measured counterclockwise from the vertical. Since the displacements and rotations are by hypothesis small, then $|\theta_i| = |y_1|$. However, $\theta_i = - y_1$ explicitly. Also, note that the force P_1 in finite elements is defined positive opposite the convention of V_1 in standard beam theory. If the two defined displacement and forces vectors are equivalent, deferring to finite element terminology, then [A] must be replaced by [A']. The difference between the two matrices is a negative sign multiplied through the second and third rows (corresponding to the redefinition of the directions of $y_i{}'$ and V_1). The result is

$$[A'] = \frac{EI}{L^3} \begin{bmatrix} 12 & 6L & -12 & 6L \\ 6L & 4L^2 & -6L & 2L^2 \\ -12 & -6L & 12 & -6L \\ 6L & 2L^2 & -6L & 4L^2 \end{bmatrix} \tag{1.59}$$

which is identical to the stiffness matrix of Equation (1.2).

Also interesting is the comparison of the right-hand side of Equation (1.2) to the right-hand side of Equation (1.45), defined by Equation (1.47). In finite elements, the loading is always lumped at the nodes. Therefore, the P_i, P_j terms contain the nodal loads due to $p(x)$ as well as the reactive shear and bending moments if the element is supported at the node. The separation of these two effects is readily apparent in Equation (1.47). The $[G]^{-1}\{B\}$ term, when multiplied out gives the finite element shape functions under the integral sign. These terms are added to the $\{F\}$ vector, which itself enters the final equation only if the beam has an external unknown reaction at the node. It is interesting to achieve this extra insight that is not always so graphic when the finite element theory is developed.

1.3 Concluding Remark

A detailed derivation of the boundary element method applied to elastic beams has been developed from the Somigliana identity. The procedure has been applied to an example cantilevered beam, and the novel features of BEM have been clearly pointed out. By manipulating the boundary element equations, the formal equivalence of the boundary element method and the finite element method in beam theory has been established. This has been done by showing that the usual symmetric matrices, typical in energy conservative finite element analysis, are achieved in the BEM beam analysis.

In Chapter 2, the more general case of the biharmonic equation in two spatial dimensions will be presented.

CHAPTER 2

TWO-DIMENSIONAL BIHARMONIC PHENOMENA

2.1 Introduction

Most of the useful analysis involving biharmonic operators involves one or two spatial dimensions. It is the purpose of this chapter to give a detailed discussion of boundary integral equation analysis applied to the two-dimensional biharmonic operator.

Two major applications of the boundary element method to two-dimensional biharmonic analysis are found in the theory of thin plates and the flow of an incompressible viscous fluid. Considerable work has been done in applying the boundary element method to the theory of thin plates. Jaswon, Maiti, and Symm (1967) developed a boundary integral equation technique for biharmonic analysis with applications in two dimensional stress problems. In their work, the biharmonic function was presented as a quadratic combination of two Laplacian functions. The resulting solution is calculated from boundary integrals involving harmonic potentials. Jaswon and Maiti (1968) extended their previous work on integral equations to the problem of clamped and simply supported plates. Other authors have presented formulations in which the biharmonic form of the fundamental solution is incorporated and applied to a variety of plate problems (Segedin and Brickell, 1968; Maiti and Chakrabarty, 1974; Altiero and Sikarskie, 1978; Stern 1979, Wu and Altiero, 1979; and Guo-Shu and Mukherjee, 1986). The approximation of the boundary in these early works was generally limited to linear variations in the geometry and a piecewise constant distribution of the biharmonic function. In most cases, the nonhomogeneous term involving the loading function was either evaluated using some form of explicit domain quadrature or separated from the numerical analysis by some change of variable.

Determining the flow field of an incompressible viscous fluid using the boundary element method was presented in a series of papers by Kelmanson, 1983(a) and 1983(b), Ingham and Kelmanson, 1984; and Hildyard et al. 1985. However, these works were limited to very slow flows which are governed adequately by the homogeneous form of the biharmonic equation. Also, the approximation of the boundary was restricted to a simple constant element formulation. If a non-zero Reynolds number flow is assumed, the governing equation becomes nonlinear and some type of iterative solution involving domain quadrature is required (Mills, 1977, and Banerjee and Butterfield, 1981).

In general there are two types of integrals required for a boundary element method solution: integrations over the surface of the problem and domain integrations involving "body force" effects or the nonhomogeneous term of the governing equation. The accuracy of the surface integrations depends greatly on the level of representation of the geometry of each boundary segment. By improving the approximation of the actual surface geometry, the accumulation of any "discretization error" is reduced. Integrations involving the body force terms over the domain are equally important in developing a formulation which produces accurate results. Originally, this type of integration was performed using a variety of volume cell quadrature schemes all requiring explicit domain discretization.

Increasing the order of both the discrete approximation of the surface geometry and the distribution of the field variables over each segment provides greater accuracy in evaluating boundary integrals. Recently, a new boundary approximation, the Overhauser element, which provides intrinsic first derivative continuity between elements in both its representation of the geometry and the variation of the function has been developed (Ortiz, 1986; Walters, 1986; Ortiz, et al, 1987).

In this work, the performance of the Overhauser element for biharmonic analysis will be compared to both a linear and a quadratic element formulation for a variety of boundary conditions and geometries. A series of analytic expressions will be derived for an isoparametric linear element and for the subparametric form of both the quadratic and Overhauser elements for the required surface integrations.

Several techniques are available which eliminate the need for explicit domain discretization when evaluating the integrations involving the nonhomogeneous terms. Domain integrations of special forms of the source function may be transformed into an equivalent series of surface integrations using the appropriate form of Green's identity. However, the evaluation of the domain integrals for a general function requires some form of numerical volume quadrature. The method presented in this work will avoid any form of explicit domain discretization and will be intrinsically sensitive to the singular nature of the integrations. The resulting formulation will reduce both the modeling and the execution time of the formulation as well as improve the accuracy of the solution at both the boundary and internal points.

The objective of this work is to develop a general boundary element formulation for the nonhomogeneous biharmonic equation of higher accuracy then affordable with earlier methods for equivalent computational effort. In doing so, several numerical improvements will be developed to increase the accuracy of the solution and reduce the execution time of the formulation. In addition, a scheme for dealing with nonhomogeneous terms that are a function of the field variables and their derivatives will be implemented which will provide an efficient way to calculate iterative and nonlinear solutions of the biharmonic equations.

2.2 Boundary Integral Equation Formulation

2.2.1 Theory

The integral equation form of the nonhomogeneous biharmonic equation may be derived several different ways. A general approach, common to many boundary element researchers, is the weighted residual technique. From this general principle, a variety of approximation schemes have developed. Some of the more widely used methods can be found in Lapidus and Pinder (1982). All weighted residual methods are similar in the respect that the unknown function is replaced by an approximation in the form of a finite linear combination of basis functions. In the finite element method, the basis functions are constructed to satisfy certain behavioral requirements over each "element" of the problem domain. The result is a polynomial form of the basis functions referred to also as a shape function or interpolation function. In a boundary element method, the finite sum approximation is represented by a combination of a shape function set and a weighting function of a particular form, referred to as the Green's function or the fundamental solution. The derivation of the fundamental solution for the biharmonic equation will be discussed later in this section.

The development of the boundary integral formulation for a non-homogeneous biharmonic equation from a weighted residual technique is not difficult. However, it is quite cumbersome, particularly in performing the integration by parts necessary to convert the domain integrals into exclusive boundary integrals. The result of this operation is to form the "inverse" problem where the biharmonic operator has been transformed from the field variable to the weighting function (Brebbia and Walker, 1980). However, the dual problem can be achieved much faster and in a more mathematically elegant fashion by using the Rayleigh-Green identity for two biharmonic functions (Jaswon and Symm, 1977). The method employed in this work will be based upon a boundary integral equation derived from the Rayleigh-Green identity.

Referring to Figure 2.1, consider the general nonhomogeneous biharmonic equation in a two-dimensional domain V,

$$\nabla^4 \psi = f(x,y) \tag{2.1}$$

The nonhomogenous function $f(x,y)$ is a known function of the spatial coordinates. In Chapter 3 of this work, the possibility of $f(x,y)$ being a function of both the coordinates and the field variable will be explored. The boundary conditions for a general biharmonic problem are of four types:

$$\psi \equiv \overline{\psi} \text{ on } S_1 \qquad\qquad \psi' \equiv \frac{\partial \overline{\psi}}{\partial n} = \overline{\psi}' \text{ on } S_2$$

$$\omega \equiv \nabla^2 \psi \equiv \overline{\omega} \text{ on } S_3 \qquad \omega' \equiv \frac{\partial(\nabla^2 \overline{\psi})}{\partial n} \equiv \overline{\omega}' \text{ on } S_4 \tag{2.2}$$

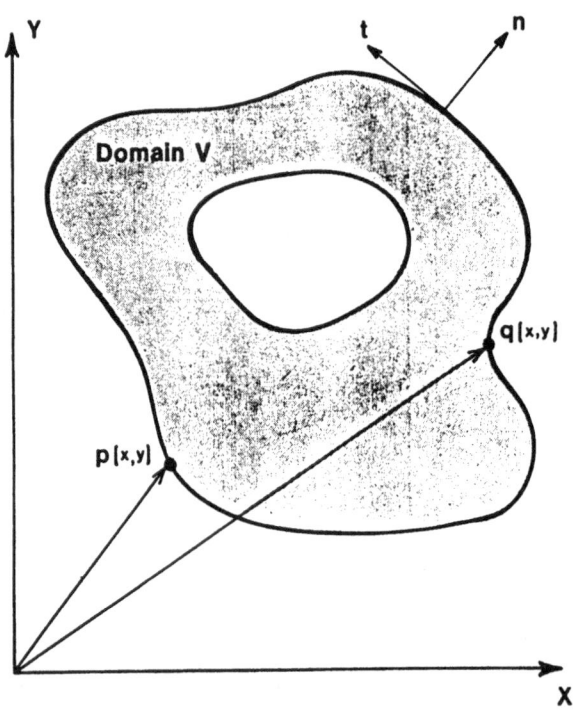

Figure 2.1   Biharmonic Problem Definition

The partial derivative with respect to n denotes the normal derivative with respect to the outward normal. For a general well-posed boundary value problem involving the biharmonic operator, two of the four types of boundary conditions are
prescribed at each point. The remaining two boundary quantities require another
functional constraint in addition to Equation (2.1). In other words, two equations are necessary in order to solve for the two remaining unknown boundary
quantities. Because of the nature of the boundary conditions it is practical to
introduce the Laplacian ω of the field function ψ explicity.

Equation (2.1) may be transformed to an equivalent set of coupled Poisson-
type equations by employing the relationship between the field variable ψ and its
Laplacian, ω :

$$\nabla^2 \psi = \omega \tag{2.3}$$

$$\nabla^2 \omega = f(x,y) \tag{2.4}$$

The result, Equation (2.4), constitutes the second functional constraint on the
biharmonic problem.

The first step in transforming Equations (2.1) and (2.4) into appropriate
integral representations is the application of the Rayleigh-Green identity for
two biharmonic functions to Equation (2.1) and Green's second identity for two
Laplacian functions to Equation (2.4). The Rayleigh-Green identity for two biharmonic functions ψ and λ which are continuous in the domain V bounded by a closed
surface S and differentiable to the fourth order in V is given as (Jaswon and
Symm, 1977)

$$\int_V (\psi \nabla^4 \lambda - \lambda \nabla^4 \psi) dV = \int_S [\psi \frac{\partial}{\partial n} (\nabla^2 \lambda) - \nabla^2 \lambda \frac{\partial \psi}{\partial n} \tag{2.5}$$
$$+ \nabla^2 \psi \frac{\partial \lambda}{\partial n} - \lambda \frac{\partial}{\partial n} (\nabla^2 \psi)] dS$$

Equation (2.5) defines the relationship between the biharmonic operator as a
domain integral and a series of surface integrations. Notice that the surface
integral terms are combinations of the two biharmonic functions ψ and λ and that

their derivatives are in a form identical to that of the above mentioned boundary conditions.

Green's second identity for two Laplacian functions ω and ϕ in a domain V bounded by a closed surface S, where the functions have continuous second derivatives, is (Jaswon and Symm, 1977)

$$\int_V (\omega\nabla^2\phi - \phi\nabla^2\omega)dV = \int_S (\omega \frac{\partial\phi}{\partial n} - \phi \frac{\partial\omega}{\partial n})dS \qquad (2.6)$$

The terms involved in the surface integrals on the right-hand side of Equation (2.6) are, like their counterparts in Equation (2.5), in the form of the previously defined boundary conditions.

An intermediate integral representation of the biharmonic equation may be accomplished by a direct and straightforward application of the Rayleigh-Green identity. The left-hand side of Equation (2.5) contains the biharmonic operator acting on both ψ and λ. The $\lambda\nabla^4\psi$ term may be viewed as the biharmonic operator acting on the field variable ψ multiplied by a weighting function λ. The second term $\psi\nabla^4\lambda$ characterizes the inverse problem, in which the biharmonic operator is acting on the weighting function. The volume integrals of Equation (2.6) contain two terms: the first term, $\phi\nabla^2\omega$, is the Laplacian of ω multiplied by a different weighting function, and the second term, $\omega\nabla^2\phi$, is the inverse relationship.

The final step in transforming Equations (2.1) and (2.4) into an integral equation form is the determination of the appropriate weighting functions. In boundary element analysis, the weighting functions are the fundamental solutions or the Green's functions for the operators in question. In general, the determination of the Green's function for a particular operator may be difficult.

Consider the vector $\vec{p}$ as the position of a variable field point where the solution is desired and the vector $\vec{q}$ as the general location of a point on the boundary or in the domain. In terms of this notation the required Green's functions are defined as the solutions to the following relationships (Brebbia, 1978):

$$\nabla^2 G_1(\vec{p},\vec{q}) = \delta(|\vec{p}-\vec{q}|) \tag{2.7}$$

$$\nabla^4 G_2(\vec{p},\vec{q}) = \delta(|\vec{p}-\vec{q}|) \tag{2.8}$$

where δ is the Dirac delta function. Solving Equations (2.7) and (2.8) defines the biharmonic and the Laplacian fundamental solutions for unbounded space:

$$G_1(\vec{p},\vec{q}) = \frac{1}{2\pi} \ln|\vec{p}-\vec{q}| \tag{2.9}$$

$$G_2(\vec{p},\vec{q}) = \frac{1}{8\pi} |\vec{p}-\vec{q}|^2 [\ln|\vec{p}-\vec{q}| - 1] \tag{2.10}$$

The integral representation of Equation (2.1) can now be obtained by using the Rayleigh-Green identity for the biharmonic function ψ and substituting G_2 for the biharmonic function λ. Applying Green's second identity to Equation (2.4) with G_1 substituted for the Laplacian function ϕ defines the integral expression of the second equation. The resulting set of coupled integral equations for a general field point are

$$\beta(\vec{p})\psi(\vec{p}) = \int_S [\psi(\vec{q})G_1'(\vec{p},\vec{q}) - \psi'(\vec{q})G_1(\vec{p},\vec{q})$$

$$+ \omega(\vec{q})G_2'(\vec{p},\vec{q}) - \omega'(\vec{q})G_2(\vec{p},\vec{q})]dS \tag{2.11}$$

$$+ \int_V f(x,y)G_2(\vec{p},\vec{q})dV$$

$$\beta(\vec{p})\omega(\vec{p}) = \int_S [\omega(\vec{q})G_1'(\vec{p},\vec{q}) - \omega'(\vec{q})G_1(\vec{p},\vec{q})]dS$$

$$+ \int_V f(x,y)G_1(\vec{p},\vec{q})dV \tag{2.12}$$

where the primes denote differentiation with respect to the outward normal of the boundary S defining a region V. The normal derivatives of each fundamental solution can be calculated in a very straight-forward manner and are defined as

$$G_1'(\vec{p},\vec{q}) = \frac{(x-x_p)n_x + (y-y_p)n_y}{2\pi|\vec{p}-\vec{q}|^2} \tag{2.13}$$

$$G_2'(\vec{p},\vec{q}) = \frac{\ln|\vec{p}-\vec{q}|^2}{8\pi} [(x-x_p)n_x + (y-y_p)n_y] \tag{2.14}$$

The value of the generalized function $\beta(\vec{p})$ shown in Equations (2.11) and (2.12) is 1.0 for a point inside the domain, some fractional value on the boundary, and is zero outside the domain (Brebbia, 1978). The solution of the two coupled integral Equations (2.11) and (2.12) requires information on the boundary for $\psi(\vec{q})$, $\psi'(\vec{q})$, $\omega(\vec{q})$, and $\omega'(\vec{q})$. Two of these quantities are defined at each boundary point $\vec{q}$ by the boundary conditions of the biharmonic problem under consideration, as shown in Figure 2.1. The remaining two quantities are determined by applying Equations (2.11) and (2.12) at points $\vec{q}$ along the boundary. Once the remaining two boundary values are determined, the values for ψ and ω may be obtained at any point within the domain.

The derivatives of ψ and ω may be calculated by differentiating the integral Equations (2.11) and (2.12) with respect to the appropriate spatial coordinate. The location of the field point where the derivatives are sought is defined by the vector $\vec{p}(x,y)$. Therefore, the spatial differential operator acts on components which are functions of $\vec{p}$ only. For example, the first derivative with respect to the x-coordinate of the functions ψ and ω are calculated as follows:

$$\frac{\partial \psi}{\partial x_p} = \int_S \left(\psi \frac{\partial G_1'}{\partial x_p} - \psi' \frac{\partial G_1}{\partial x_p} + \omega \frac{\partial G_2'}{\partial x_p} - \omega' \frac{\partial G_2}{\partial x_p} \right) dS$$

$$+ \int_V f(x,y) \frac{\partial G_2}{\partial x_p} dV \tag{2.15}$$

$$\frac{\partial \omega}{\partial x_p} = \int_S \left(\omega \frac{\partial G_1'}{\partial x_p} - \omega' \frac{\partial G_1}{\partial x_p} \right) dS + \int_V f(x,y) \frac{\partial G_1}{\partial x_p} dV \tag{2.16}$$

where the derivative of the Green's functions G_1, G_1', G_2, and G_2' with respect to the x-coordinate are calculated as

$$\frac{\partial G_1}{\partial x_p} = \frac{1}{2\pi} \frac{(x-x_p)}{|\vec{p}-\vec{q}|^2} \tag{2.17}$$

$$\frac{\partial G_1'}{\partial x_p} = -\frac{1}{2\pi} \left(\frac{2[(x-x_p)^2 n_x + (x-x_p)(y-y_p)n_y]}{|\vec{p}-\vec{q}|^4} - \frac{n_x}{|\vec{p}-\vec{q}|^2} \right) \tag{2.18}$$

$$\frac{\partial G_2}{\partial x_p} = \frac{1}{8\pi} [(x-x_p)(\ln|\vec{p}-\vec{q}|^2 - 1)] \qquad (2.19)$$

$$\frac{\partial G_2'}{\partial x_p} = \frac{1}{8\pi} (\frac{2(x-x_p)}{|\vec{p}-\vec{q}|^2} [(x-x_p)n_p + (y-y_p)n_y]$$

$$+ n_x [\ln|\vec{p}-\vec{q}|^2 - 1]) \qquad (2.20)$$

In a similar manner, the first derivative of ψ and ω with respect to the y-coordinate may be determined from the following expressions:

$$\frac{\partial \psi}{\partial y_p} = \int_s (\psi \frac{\partial G_1'}{\partial y_p} - \psi' \frac{\partial G_1}{\partial y_p} + \omega \frac{\partial G_2'}{\partial y_p} - \omega' \frac{\partial G_2}{\partial y_p})dS$$

$$+ \int_V f(x,y) \frac{\partial G_2}{\partial y_p} dV \qquad (2.21)$$

$$\frac{\partial \omega}{\partial y_p} = \int_s (\omega \frac{\partial G_1'}{\partial y_p} - \omega' \frac{\partial G_1}{\partial y_p})dS + \int_V f(x,y) \frac{\partial G_1}{\partial y_p} dV \qquad (2.22)$$

where the derivative of the Green's functions G_1, G_1', G_2, and G_2' with respect to the y-coordinate are calculated as

$$\frac{\partial G_1}{\partial y_p} = \frac{1}{2\pi} \frac{(x-x_p)}{|\vec{p}-\vec{q}|^2} \qquad (2.23)$$

$$\frac{\partial G_1'}{\partial y_p} = - \frac{1}{2\pi} (\frac{2[(x-x_p)(y-y_p)n_x + (y-y_p)^2 n_y]}{|\vec{p}-\vec{q}|^4} - \frac{n_y}{|\vec{p}-\vec{q}|^2}) \qquad (2.24)$$

$$\frac{\partial G_2}{\partial y_p} = \frac{1}{8\pi} [(y-y_p)(\ln|\vec{p}-\vec{q}|^2 - 1)] \qquad (2.25)$$

$$\frac{\partial G_2'}{\partial y_p} = \frac{1}{8\pi} (\frac{2(y-y_p)}{|\vec{p}-\vec{q}|^2} [(x-x_p)n_x + (y-y_p)n_y]$$

$$+ n_y [\ln|\vec{p}-\vec{q}|^2 - 1]) \qquad (2.26)$$

The value of any order spatial derivative in the domain interior is a function of the same four boundary quantities, $\psi(\vec{q})$, $\psi'(\vec{q})$, $\omega(\vec{q})$, and $\omega'(\vec{q})$, and the nonhomogeneous function $f(x,y)$ that are used in the calculation of field variables ψ and ω at any point. The calculation of any order derivative with respect to any spatial coordinate at an internal point may be accomplished by determining the appropriate derivative forms of the Green's functions and substituting them into Equations (2.11) and (2.12).

2.2.2 Boundary Element Formulation

The term "boundary elements" was first used in association with the boundary integral equation method to indicate a technique whereby the boundary of a problem domain is subdivided into a series of finite elements over which a field variable is approximated (Brebbia, 1978). The obvious advantage of boundary elements over more traditional methods such as finite element and finite difference techniques is a reduction in the order of the dimensionality of the problem by one. A general multi-dimensional boundary value problem may be approximated through a series of surface integrations rather than a set of domain integrations. The resulting integral equations require information on the geometry and the field variables at points along the problem surface, thereby reducing the amount of information necessary to accurately describe the physical problem.

The first approximation in the boundary element method is the discretization of the problem surface into a series of elements. The behavior of the field variables ψ and ω and their normal derivatives ψ' and ω' in Equations (2.11) and (2.12) over each boundary element is characterized by an assumed interpolation function. As in finite element methods, these interpolation functions or shape functions can be of many different forms and result in varying degrees of accuracy for the field variables and the surface geometry. If the shape function defining the distribution of the field variable and the geometry over an element are the same, the element is called isoparametric. An element where the variation of the geometry is defined by a lower order shape function than that used to describe the field variable is termed subparametric. A third element type is superparametric, in which the order of the shape function defining the geometry is higher than that used to distribute the field variable over the element. Advantages and disadvantages associated with each element type will be discussed in Chapter 3.

By defining the interpolation function as a shape function set {N}, a column vector, the distribution of ψ, ω, ψ', and ω' over each element may be established as

$$\psi = <\psi> \{N\} \qquad \psi' = <\psi'> \{N\}$$
$$\omega = <\omega> \{N\} \qquad \omega' = <\omega'> \{N\} \tag{2.27}$$

where $<\psi>$, $<\omega>$, $<\psi'>$, and $<\omega'>$ are row vectors containing the discrete values of ψ, ω, ψ', and ω' respectively at nodes defining each element. Throughout this text, the symbols $<*>$ will denote a row vector, $\{*\}$ will denote a column vector, and $[*]$ will represent a general rectangular matrix. Substituting these approximations into Equations (2.11) and (2.12) results in the following discrete expressions for the governing set of coupled integral equations

$$\sum_{j=1}^{n} \left(\int_{S_j} [<N>G_1'(\vec{p},\vec{q})\{\psi\} - <N>G_1(\vec{p},\vec{q})\{\psi'\} \right.$$

$$+ <N>G_2'(\vec{p},\vec{q})\{\omega\} - <N>G_2(\vec{p},\vec{q})\{\omega'\}]dS_j \left. \right)$$

$$+ \int_V F(x,y)G_2(\vec{p},\vec{q})dV - \beta(\vec{p})\psi(\vec{p}) = 0 \tag{2.28}$$

$$\sum_{j=1}^{n} \left(\int_{S_j} [<N>G_1'(\vec{p},\vec{q})\{\omega\} - <N>G_1(\vec{p},\vec{q})\{\omega'\}]dS_j \right)$$

$$+ \int_V f(x,y)G_1(\vec{p},\vec{q})dV - \beta(\vec{p})\omega(\vec{p}) = 0 \tag{2.29}$$

where the summation is over n elements that define the boundary. The integrands of Equations (2.28) and (2.29) may be rewritten by introducing the following terms

$$H_{ij} = \frac{1}{2\pi} \int_{S_j} <N> \ln' |\vec{q}_i - \vec{q}| dS_j - \beta(\vec{p})\delta_{ij} \tag{2.30}$$

where $\delta_{ij} \equiv$ Kronecker delta;

$$G_{ij} = \frac{1}{2\pi} \int_{S_j} <N> \ln |\vec{q}_i - \vec{q}| dS_j \tag{2.31}$$

$$L_{ij} = \frac{1}{8\pi} \int_{S_j} <N> (|\vec{q}_i - \vec{q}|^2 [\ln |\vec{q}_i - \vec{q}| - 1])' dS_j \tag{2.32}$$

$$K_{ij} = \frac{1}{8\pi} \int_{S_j} <N> (|\vec{q}_i - \vec{q}|^2 [\ln |\vec{q}_i - \vec{q}| - 1]) dS_j \tag{2.33}$$

$$B1_i = \frac{1}{8\pi} \int_V f(x,y) (|\vec{q}_i - \vec{q}|^2 [\ln |\vec{q}_i - \vec{q}| - 1]) dV \tag{2.34}$$

$$B2_i = \frac{1}{2\pi} \int_V f(x,y) \ln |\vec{q}_i - \vec{q}| dV \tag{2.35}$$

The integrals in Equations (2.30) - (2.35) may be evaluated analytically for linear isoparametric and higher order subparametric elements, and for certain forms of the nonhomogeneous function $f(x,y)$. The exact evaluations avoid the error introduced by numerical quadrature schemes and generally decrease computational time while increasing the accuracy of the integration. Unless specially formulated, most numerical quadrature schemes become inaccurate at small values of $|\vec{q}_i - \vec{q}|$. This type of error is especially evident at internal point calculations very close to the boundary. The reader is referred to Chapter 3 for the details of the analysis involved in obtaining the exact expressions for Equations (2.30) - (2.35) for the above mentioned elements.

Substituting Equations (2.30) - (2.35) into Equations (2.28) and (2.29) reduces the problem formulation to a coupled set of vector equations with the form

$$[H]\{\psi\} + [G]\{\psi'\} + [L]\{\omega\} + [K]\{\omega'\} = \{B1\} \tag{2.36}$$

$$[H]\{\omega\} + [G]\{\omega'\} = \{B2\} \tag{2.37}$$

where the column vectors $\{\psi\}$, $\{\psi'\}$, $\{\omega\}$, and $\{\omega'\}$ represent the values of ψ, ψ', ω, ω' at each node. The diagonal terms of the [H] matrix contain the constant $\beta(\vec{p})$, but there is no need to explicitly perform the integrations to obtain this value. The diagonal terms of [H] may be calculated from the homogeneous form of Equation (2.37) by applying the fact that when a uniform potential, say unity, is applied over the entire boundary, the normal derivatives on the boundary must be zero everywhere. Therefore, Equation (2.37) becomes

$$[H] \{1\} = \{0\} \tag{2.38}$$

This equation states that the sum of the elements in each row of the [H] matrix must be zero. Therefore, the diagonal term of a row in [H] is the negative of the sum of all nondiagonal terms of that row (Brebbia, 1978).

Equations (2.36) and (2.37) may be rewritten in a more compact form as a single vector equation by combining terms involving the functions ψ and ω into one matrix, and the normal derivative terms ψ' and ω' into a second matrix:

$$\begin{bmatrix} [H] & [L] \\ 0 & [H] \end{bmatrix} \begin{Bmatrix} \{\psi\} \\ \{\omega\} \end{Bmatrix} = \begin{bmatrix} [G] & [K] \\ 0 & [G] \end{bmatrix} \begin{Bmatrix} \{\psi'\} \\ \{\omega'\} \end{Bmatrix} + \begin{Bmatrix} \{B1\} \\ \{B2\} \end{Bmatrix} \tag{2.39}$$

At any point on the boundary at least two of the four quantities ψ, ω, ψ', and ω' are specified. Depending on the combination of boundary conditions prescribed at a discrete point, the columns of the matrices in Equation (2.39) may be rearranged such that all the unknown boundary quantities are on one side of the equation. The result is the construction of a matrix equation of the form

$$[A]\{x\} = \{B\} \tag{2.40}$$

where [A] is a nonsymmetric matrix, $\{x\}$ is a column vector of unknown boundary quantities, and $\{B\}$ is a column vector calculated from the prescribed boundary conditions and their appropriated matrix components combined with the domain integral terms.

2.3 Summary and Concluding Remarks

A boundary element formulation for the nonhomogeneous biharmonic equation has been presented in this chapter. Boundary element analysis has many appealing advantages over the more traditional domain type formulations such as the finite element and finite difference techniques. Since only the boundary surface is modeled, the dimensionality of the problem is reduced by one. Consequently, both the input information necessary to define the problem and the simultaneous equations required for a solution are generally reduced. Another advantage is that for certain types of problems, the accuracy and consistency of the results from a boundary element solution can be considerably better than those obtained from either a finite element or a finite difference method (Connor and Brebbia, 1986).

The next chapter will introduce several different types of elements for use in the boundary discretization and the derivation of some analytical expressions for the integrals defined in Equations (2.30) - (2.33). Also presented are various techniques to calculate the domain integrals of Equations (2.34) and (2.35) which avoid the disadvantages associated with domain cell methods. The resulting numerical analysis procedure will be capable of solving the nonhomogeneous problem as easily as the homogeneous form.

CHAPTER 3

ELEMENT TYPES FOR BOUNDARY DISCRETIZATION

3.1 Introduction

In Chapter 2, a boundary integral equation representation of the nonhomogeneous biharmonic equation was formulated in terms of the field variables ψ and ω,
their normal derivatives ψ' and ω', and the nonhomogeneous function $f(x,y)$. An
assumed set of shape functions were defined over each boundary element which
characterized the distribution of the four boundary quantities along the surface. An example was presented using the simplest possible boundary representation a so-called "constant" element. In this chapter, a detailed discussion of
the more sophisticated element types is presented. It is with these true finite
element shape functions that the name "boundary elements" is warranted. Three
isoparametric shape function sets are developed: a two node linear element, a
three node quadratic element, and a recently developed two node Overhauser element defined by four nodes. In addition, for rectilinear geometries a subparametric version of both the quadratic and Overhauser elements will be defined.

Analytic expressions for the integrations of Equations (2.30) - (2.33) are
derived for an isoparametric linear element and the subparametric form of both
the quadratic and the Overhauser elements. For all other cases, a general numerical form of the integrations of Equations (2.30) - (2.33) are presented for
quadratic and Overhauser elements. The "constant" element case is deliberately
omitted since it has been analyzed exhaustively in previous work (i.e. Kelmanson,
1983(a), 1983(b)).

3.2 Linear Elements

3.2.1 Isoparametric Linear Elements

The boundary will be divided into n straight line segments, and a linear distribution of the boundary quantities over each element will be assumed. A general linear element defined by two endpoints, node "i" and node "j", as shown in Figure 3.1, can be transformed into a one dimensional space of a single parameter, t. The resulting isoparametric element is C_0 continuous in each of the four boundary quantities ψ, ω, ψ', and ω'. The values of any one of the boundary quantities at any point t along the element is defined in terms of their discrete nodal values and a shape function set <N>. For example, the linear distribution of the field variables ψ and ω is given as

$$\psi(t) = N_i \psi_i + N_j \psi_j = \langle N_i \; N_j \rangle \begin{Bmatrix} \psi_i \\ \psi_j \end{Bmatrix}$$

$$\omega(t) = N_i \omega_i + N_j \omega_j = \langle N_i \; N_j \rangle \begin{Bmatrix} \omega_i \\ \omega_j \end{Bmatrix}$$

(3.1)

The shape functions N_i and N_j are

$$N_i = t \qquad\qquad N_j = 1-t \tag{3.2}$$

The form of the integrals in Equations (2.30) - (2.33) are transformed into the parameter space reducing the order of the integration by one. Using the shape functions defined in Equation (3.2), the variation over an element of the two-dimensional coordinates x and y can be written as

$$x = x_i(1-t) + x_j t \qquad\qquad y = y_i(1-t) + y_j t \tag{3.3}$$

The transformation of the differential length dS is accomplished by using the following simple one dimensional Jacobian:

$$|J| = \sqrt{\left(\frac{dx}{dt}\right)^2 + \left(\frac{dy}{dt}\right)^2} = \frac{dS}{dt} \tag{3.4}$$

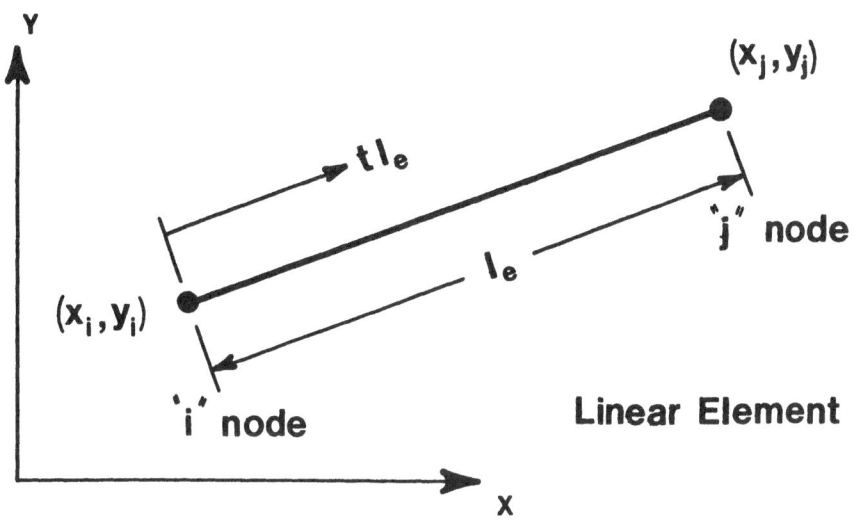

Linear Element

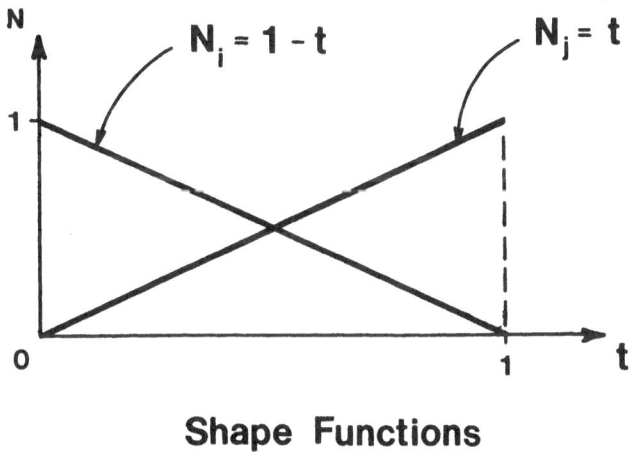

Shape Functions

Figure 3.1 Linear Element Nomenclature

$$dS = |J|dt = l_e dt \qquad\qquad l_e \equiv \text{Element Length} \qquad\qquad (3.5)$$

The argument of the fundamental solution for each integrand of Equations (2.30) – (2.33) are of the form $|\vec{q}_p - \vec{q}|$. For a linear element, the argument may be replaced using the following relationships:

$$|\vec{q}_i - \vec{q}|^2 = (x(t) - x_p)^2 + (y(t) - y_p)^2 \qquad\qquad (3.6)$$

where (x_p, y_p) are the coordinates which locate a variable field point. Expanding Equation (3.6) in terms of the parameter t results in the following expression for $|\vec{q}_p - \vec{q}|^2$:

$$|\vec{q}_p - \vec{q}|^2 = ((x_j - x_i)t + x_i - x_p)^2 + ((y_j - y_i) + y_i - y_p)^2 \qquad\qquad (3.7)$$

Equation (3.7) may be rewritten as

$$|\vec{q}_p - \vec{q}|^2 = At^2 + Bt + C \equiv X \qquad\qquad (3.8)$$

where the constants A, B, and C, graphically shown in Figure 3.2, are defined as

$$A = A_x^2 + A_y^2 \qquad B = 2(A_x B_x + A_y B_y) \qquad C = B_x^2 + B_y^2$$

$$A_x = x_j - x_i \qquad B_x = x_i - x_p \qquad\qquad (3.9)$$

$$A_y = y_j - y_i \qquad B_y = y_i - y_p$$

Substituting the above expression into the integrals of Equations (2.30) – (2.33) results in the following parametrized forms:

$$H_{pe} = \frac{l_e}{2\pi} \int_0^1 <t, 1-t> \frac{Dt+E}{X} \{\psi_e\} dt - \beta(\vec{p})\delta_{pe} \qquad\qquad (3.10)$$

$$G_{pe} = \frac{l_e}{4\pi} \int_0^1 <t, 1-t> \ln X \{\psi_e'\} dt \qquad\qquad (3.11)$$

$$L_{pe} = \frac{l_e}{8\pi} \int_0^1 <t, 1-t> (\ln X - 1)(Dt+E)\{\omega_e\} dt \qquad\qquad (3.12)$$

$$K_{pe} = \frac{1}{16\pi} \int_0^1 <t, \ 1-t> X(\ln X - 2) \{\omega_e'\} dt \tag{3.13}$$

where constants D and E are defined as

$$D = A_x n_x + A_y n_y \equiv 0 \qquad E = B_x n_x + B_y n_y \tag{3.14}$$

For the linear element under consideration, the constant D (which is the dot pro-
duct of the vector $\vec{A}$ defining the length of the element and the unit normal vector, as shown in Figure 3.2) is identically zero.

The integrands of the one dimensional integrations defined in Equations
(3.10) - (3.13) are combinations of the functions $t^n \ln X$, t^n / X, and t^n all of
which may be evaluated analytically. Therefore, the following integration table
may be compiled:

$$I_0 = \int_0^1 \frac{dt}{X} = \frac{2\gamma}{(4AC - B^2)^{1/2}} \qquad 4AC - B^2 > 0 \tag{3.15}$$

$$= \frac{2}{B(1 + B/2C)} \qquad 4AC - B^2 = 0$$

where the angle γ is defined in Figure 3.2:

$$I_1 = \int_0^1 \frac{t}{X} dt = (\ln(A+B+C) - \ln C - B I_0)/2A \tag{3.16}$$

$$I_2 = \int_0^1 \frac{t^2}{X} dt = (1 - B I_1 - C I_0)/A \tag{3.17}$$

$$I_3 = \int_0^1 \frac{t^3}{X} dt = (\tfrac{1}{2} - B I_2 - C I_1)/A \tag{3.18}$$

$$I_4 = \int_0^1 \frac{t^4}{X} dt = (\tfrac{1}{3} - B I_3 - C I_2)/A \tag{3.19}$$

$$I_5 = \int_0^1 \frac{t^5}{X} dt = (\tfrac{1}{4} - B I_4 - C I_3)/A \tag{3.20}$$

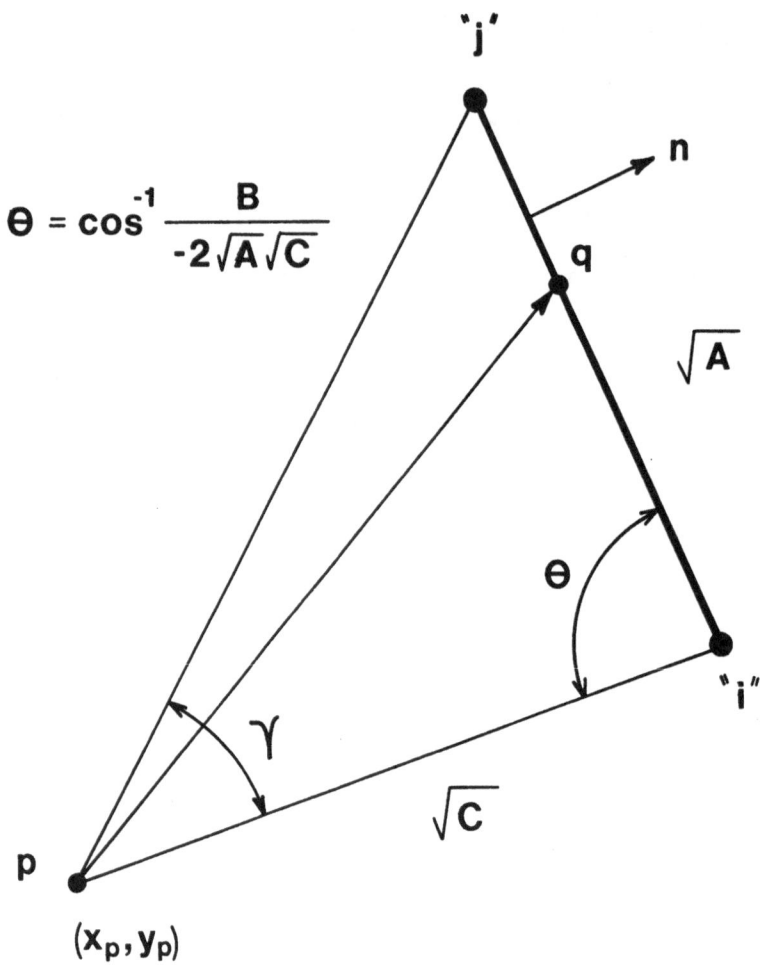

Figure 3.2   Definitions for Exact Analysis of Linear Elements

Analytic expressions for the integration of terms of the form $t^n \ln X$ may be listed as

$$L_0 = \int_0^1 \ln X \, dt = \ln(A+B+C) - 2AI_2 - BI_1 \qquad (3.21)$$

$$L_1 = \int_0^1 t \ln X \, dt = \frac{1}{2} \left(\ln(A+B+C) - 2AI_3 - BI_2 \right) \qquad (3.22)$$

$$L_2 = \int_0^1 t^2 \ln X \, dt = \frac{1}{3} \left(\ln(A+B+C) - 2AI_4 - BI_3 \right) \qquad (3.23)$$

$$L_3 = \int_0^1 t^3 \ln X \, dt = \frac{1}{4} \left(\ln(A+B+C) - 2AI_5 - BI_4 \right) \qquad (3.24)$$

Each of the parametrized integrations in Equations (3.10) - (3.13) have individual values associated with each discrete end node of the linear element being integrated. Therefore, two values are determined for each of the four integrations. In terms of the element constants A, B, C, and E, and the integration table defined previously, the exact integrations for the expression in Equations (3.10) - (3.13) may be defined as

$$H_{pe} = \frac{1_e}{2\pi} \left(H_i \psi_i + H_j \psi_j \right) \qquad (3.25)$$

$$G_{pe} = \frac{1_e}{4\pi} \left(G_i \psi_i' + G_j \psi_j' \right) \qquad (3.26)$$

$$L_{pe} = \frac{1_e}{8\pi} \left(L_i \omega_i + L_j \omega_j \right) \qquad (3.27)$$

$$K_{pe} = \frac{1_e}{16\pi} \left(K_i \omega_i' + K_j \omega_j' \right) \qquad (3.28)$$

where

$$H_i = E(I_0 - I_1) \qquad\qquad H_j = EI_1 \qquad (3.29)$$

$$G_i = L_0 - L_1 \qquad\qquad G_j = L_1 \qquad (3.30)$$

$$L_i = E\left(L_0 - L_1 - \frac{1}{2}\right) \qquad\qquad L_j = E\left(L_1 - \frac{1}{2}\right) \qquad (3.31)$$

$$K_j = AL_3 + BL_2 + CL_1 - 2\left(\frac{A}{4} + \frac{B}{3} + \frac{C}{2} \right)$$

$$K_i = AL_2 + BL_1 + CL_0 - 2\left(\frac{A}{3} + \frac{B}{2} + C \right) - K_j$$

(3.32)

The preceding analytical expressions for the integrations required for bi-harmonic analysis using linear elements have many advantages. The source point (x_p, y_p) may be at any location, even occupy a point on the element itself without loss of generality. Normally, if numerical quadrature were used, special care would have to be taken if the source point was a member of or colinear with the element to handle the singular part of the integral. Also, depending on discrete mesh size, values of the field variables calculated by numerical quadrature at points very near an element where the natural logarithm terms of the fundamental solutions approach their singular point may be inconsistent and inaccurate. The presented exact analysis implicitly handles the singular terms without any adjustment in the formulation. This is particularly important in calculating the system matrices and in accurately evaluating values of the field variables at internal points near the boundary. Unlike numerical procedures, the exact formulation is not iterative by nature, and therefore considerably reduces the time required to formulate the system matrices and calculate internal points.

A linear element formulation for the required integrations using analytic expressions has obvious advantages over the more commonly used subparametric constant element. Not only is the order of the approximation increased by one, from a constant to a linear function, but more importantly, the linear shape functions provide C_0 continuity for all four boundary quantities ψ, ω, ψ', and ω' between elements.

The major disadvantage intrinsic to linear elements is their inability to accurately describe complex geometries and rapidly varying functions. Higher order elements are required to better represent the geometry of the domain and consequently increase the accuracy of the approximation.

3.2.2 Superparametric Linear Elements - Constant Elements

A type of "constant" element boundary approximation may be thought of as a superparametric form of the linear element, wherein the variation of the field variables over each boundary segment is considered uniform. The resulting boundary element approximation provides a discontinuous representation of the field variables between elements while maintaining C_0 continuity in the geometry.

A parametric form of each of the integrations defined in Equations (2.30) - (2.33) may be written for a constant element by slightly modifying the parametric linear element equations (Equations (3.10) - (3.13)). The value of any one of the boundary quantities ψ, ψ', ω, and ω' at point along the element is defined in terms of a discrete nodal value, generally located at the element center, and the "shape function" $\langle N \rangle = 1$. However, the variation over an element of the two-dimensional coordinates (x,y) is the same as in their linear element counterparts, Equation (3.2). The parametric form of the argument $|\vec{q}_p - \vec{q}|$ is given by Equation (3.8) using the same set of element constants defined in Equations (3.9). Substituting these relationships into the integrals of Equations (2.30) - (2.33) results in the following parameterized forms

$$H_{pi} = \frac{l_e}{2\pi} \int_0^1 \frac{E}{X} \{\psi_i\} \, dt - \beta(\vec{p})\delta_{pi} \tag{3.33}$$

$$G_{pi} = \frac{l_e}{4\pi} \int_0^1 \ln X \{\psi_i'\} \, dt \tag{3.34}$$

$$L_{pi} = \frac{l_e}{8\pi} \int_0^1 E(\ln X - 1) \{\omega_i\} \, dt \tag{3.35}$$

$$K_{pi} = \frac{l_e}{16\pi} \int_0^1 X(\ln X - 2) \{\omega_i'\} \, dt \tag{3.36}$$

where the constant E is defined in Equation (3.14)

The integrands of the parameterized integrations defined in Equations (3.33) - (3.36) are combinations of the functions $t^n \ln X$ and t^n/X. Analytical expressions for each of these functions have been previously defined in the integration table of Equations (3.15) - (3.24). Each of the four integrations has a single value associated with the discrete center node of the constant element being integrated. Therefore, in terms of the element constants A, B, C, E, and the analytical expression compiled in the integration table, the exact integrations for Equations (3.33) - (3.36) may be defined as

$$H_{pi} = \frac{l_i}{2\pi} E \ I_0 \ \{\psi_i\} \tag{3.37}$$

$$G_{pi} = \frac{l_i}{4\pi} L_0 \ \{\psi_i'\} \tag{3.38}$$

$$L_{pi} = \frac{l_i}{8\pi} E \ (L_0 - 1) \ \{\omega_i\} \tag{3.39}$$

$$K_{pi} = \frac{l_i}{16\pi} (\ AL_2 + BL_1 + CL_0 - 2(\frac{A}{3} + \frac{B}{2} + C \ )) \ \{\omega_i'\} \tag{3.40}$$

The main purpose of presenting the constant element formulation in this text is twofold. First, the constant element (non-parametric form) has received considerable attention in the published literature and has been analyzed exhaustively in previous work (Kelmanson 1983(a), 1983(b)). Secondly, the constant element formulation provides the simplest means of presenting the explicit details of compiling and solving the coupled set of governing equations defined in Equations (2.36) and (2.37). In the next section of this chapter, an example using a constant element formulation will be used to demonstrate the general solution procedure utilized by the computer program presented in Chapter 6.

3.2.3 Example - Creeping Flow Between Parallel Plates

In this example, the calculation of the individual components of Equations (2.36) and (2.37) using constant elements will be presented in explicit detail for instructive purposes. This type of problem is most helpful in understanding the general concepts necessary to formulate and solve the resulting set of equations.

NO SLIP BOUNDARY CONDITION

$$\psi = 1 \ , \ \frac{\partial \psi}{\partial n} = 0$$

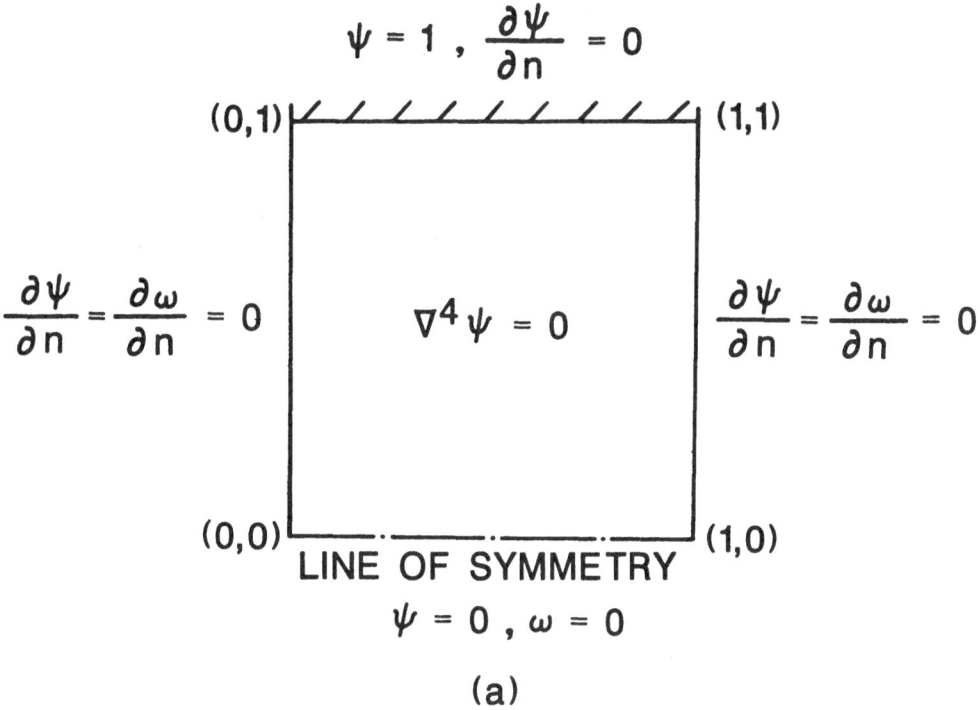

$(0,1)$ $(1,1)$

$$\frac{\partial \psi}{\partial n} = \frac{\partial \omega}{\partial n} = 0 \qquad \nabla^4 \psi = 0 \qquad \frac{\partial \psi}{\partial n} = \frac{\partial \omega}{\partial n} = 0$$

$(0,0)$ $(1,0)$

LINE OF SYMMETRY

$$\psi = 0 \ , \ \omega = 0$$

(a)

BOUNDARY ELEMENT MESH

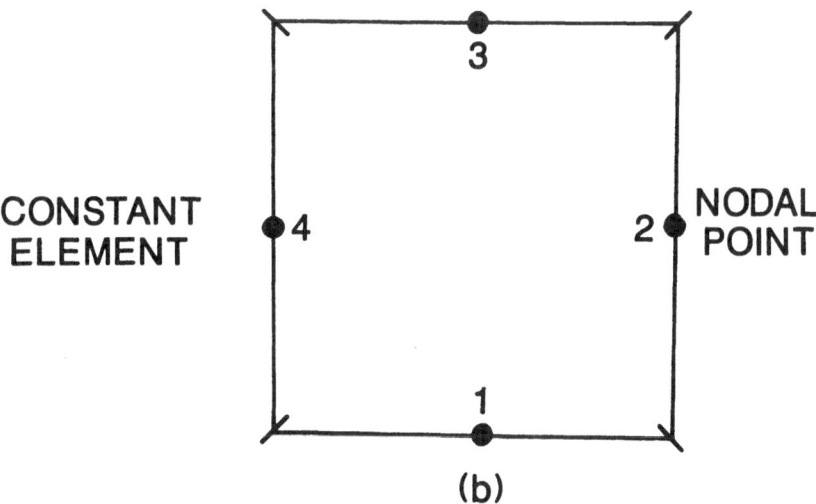

CONSTANT ELEMENT

NODAL POINT

(b)

Figure 3.3   Problem Definition and Boundary Discretization for Flow Between Two Parallel Plates.

Consider the problem of creeping flow between two parallel plates. Due to symmetry about the centerline, the problem geometry is reduced to that shown in Figure 3.3(a). The governing equation for the flow field in terms of the stream function ψ is given as:

$$\nabla^4\psi = 0 \qquad\qquad (3.41)$$

The boundary will be divided into four straight line segments, and a constant value of each of the four boundary quantities will be assumed over each segmental element. The resulting constant element boundary approximation is shown in Figure 3.3(b).

Using the analytical expression defined in Equations (3.37) - (3.40) for a constant element, the matrix components of Equation (2.39) are calculated and given as:

$$[H] = -1/2\pi \begin{bmatrix} 3.14159 & -1.10715 & -0.92730 & -1.10715 \\ -1.10715 & 3.14159 & -1.10715 & -0.92730 \\ -0.92730 & -1.10715 & 3.14159 & -1.10715 \\ -1.10715 & -0.92730 & -1.10715 & 3.14159 \end{bmatrix} \qquad (3.42)$$

$$[G] = -1/2\pi \begin{bmatrix} 1.69315 & 0.33485 & -0.03887 & 0.33485 \\ 0.33485 & 1.69315 & 0.33485 & -0.03887 \\ -0.03887 & 0.33485 & 1.69315 & 0.33485 \\ 0.33485 & -0.03887 & 0.33485 & 1.69315 \end{bmatrix} \qquad (3.43)$$

$$[L] = -1/2\pi \begin{bmatrix} 0.00000 & 0.20871 & 0.23057 & 0.20871 \\ 0.20871 & 0.00000 & 0.20871 & 0.23057 \\ 0.23057 & 0.20871 & 0.00000 & 0.20871 \\ 0.20871 & 0.23057 & 0.20871 & 0.00000 \end{bmatrix} \qquad (3.44)$$

$$[K] = -1/2\pi \begin{bmatrix} 0.04222 & 0.17594 & 0.25968 & 0.17594 \\ 0.17594 & 0.04222 & 0.17594 & 0.25968 \\ 0.25968 & 0.17594 & 0.04222 & 0.17594 \\ 0.17594 & 0.25968 & 0.17594 & 0.04222 \end{bmatrix} \tag{3.45}$$

Substituting the above matrix components into the single vector equation defined in Equation (2.39) results in:

$$\begin{bmatrix} 3.142 & -1.107 & -0.927 & -1.107 & 0.000 & 0.209 & 0.231 & 0.209 \\ -1.107 & 3.142 & -1.107 & -0.927 & 0.209 & 0.000 & 0.209 & 0.231 \\ -0.927 & -1.107 & 3.142 & -1.107 & 0.231 & 0.209 & 0.000 & 0.209 \\ -1.107 & -0.927 & -1.107 & 3.142 & 0.209 & 0.231 & 0.209 & 0.000 \\ 0.000 & 0.000 & 0.000 & 0.000 & 3.142 & -1.107 & -0.927 & -1.107 \\ 0.000 & 0.000 & 0.000 & 0.000 & -1.107 & 3.142 & -1.107 & -0.927 \\ 0.000 & 0.000 & 0.000 & 0.000 & -0.927 & -1.107 & 3.142 & -1.107 \\ 0.000 & 0.000 & 0.000 & 0.000 & -1.107 & -0.927 & -1.107 & 3.142 \end{bmatrix} \begin{Bmatrix} \psi_1 \\ \psi_2 \\ \psi_3 \\ \psi_4 \\ \omega_1 \\ \omega_2 \\ \omega_3 \\ \omega_4 \end{Bmatrix}$$

$$\tag{3.46}$$

$$= \begin{bmatrix} 1.693 & 0.335 & -0.039 & 0.335 & 0.042 & 0.176 & 0.260 & 0.176 \\ 0.335 & 1.693 & 0.335 & -0.039 & 0.176 & 0.042 & 0.176 & 0.260 \\ -0.039 & 0.335 & 1.693 & 0.335 & 0.260 & 0.176 & 0.042 & 0.176 \\ 0.335 & -0.039 & 0.335 & 1.693 & 0.176 & 0.260 & 0.176 & 0.042 \\ 0.000 & 0.000 & 0.000 & 0.000 & 1.693 & 0.335 & -0.039 & 0.335 \\ 0.000 & 0.000 & 0.000 & 0.000 & 0.335 & 1.693 & 0.335 & -0.039 \\ 0.000 & 0.000 & 0.000 & 0.000 & -0.039 & 0.335 & 1.693 & 0.335 \\ 0.000 & 0.000 & 0.000 & 0.000 & 0.335 & -0.039 & 0.335 & 1.693 \end{bmatrix} \begin{Bmatrix} \psi'_1 \\ \psi'_2 \\ \psi'_3 \\ \psi'_4 \\ \omega'_1 \\ \omega'_2 \\ \omega'_3 \\ \omega'_4 \end{Bmatrix}$$

Applying the boundary conditions defined in Figure 3.3(a) to Equation (3.46) and rearranging the columns of the matrices such that the unknowns are on the left side of the equation and the known quantities are on the right hand side of the equation results in:

$$
\begin{bmatrix}
1.693 & 1.107 & -0.231 & 1.107 & 0.042 & -0.209 & 0.260 & -0.209 \\
0.335 & -3.142 & -0.209 & 0.927 & 0.176 & 0.000 & 0.176 & -0.231 \\
-0.039 & 1.107 & 0.000 & 1.107 & 0.260 & -0.209 & 0.042 & -0.209 \\
0.335 & 0.927 & -0.209 & -3.142 & 0.176 & -0.231 & 0.176 & 0.000 \\
0.000 & 0.000 & 0.927 & 0.000 & 1.693 & 1.107 & -0.039 & 1.107 \\
0.000 & 0.000 & 1.107 & 0.000 & 0.335 & -3.142 & 0.335 & 0.927 \\
0.000 & 0.000 & -3.142 & 0.000 & -0.039 & 1.107 & 1.693 & 1.107 \\
0.000 & 0.000 & 1.107 & 0.000 & 0.335 & 0.927 & 0.335 & -3.142
\end{bmatrix}
\begin{Bmatrix}
\psi'_1 \\
\psi_2 \\
\omega_3 \\
\psi_4 \\
\omega'_1 \\
\omega_2 \\
\omega'_3 \\
\omega_4
\end{Bmatrix}
$$

$$
=
\begin{Bmatrix}
-0.92730 \\
-1.10715 \\
3.14159 \\
-1.10715 \\
0.00000 \\
0.00000 \\
0.00000 \\
0.00000
\end{Bmatrix}
\tag{3.47}
$$

The system of equations is then solved using Gaussian elimination. A comparison of the results for the unknown boundary quantities obtained from Equation (3.47) and the exact solution follows:

BEM   Exact

$$
\begin{Bmatrix} \psi'_1 \\ \psi_2 \\ \omega_3 \\ \psi_4 \\ \omega'_1 \\ \omega_2 \\ \omega'_3 \\ \omega_4 \end{Bmatrix}
=
\begin{Bmatrix} -1.820 \\ 0.702 \\ -3.265 \\ 0.702 \\ 3.835 \\ -1.632 \\ -3.835 \\ -1.632 \end{Bmatrix}
\quad
\begin{Bmatrix} -1.500 \\ 0.687 \\ -3.000 \\ 0.687 \\ 3.000 \\ -1.500 \\ -3.000 \\ -1.500 \end{Bmatrix}
\qquad (3.48)
$$

The importance of this example lies not in the accuracy of the results of the simple constant element formulation, but rather in the concepts used to construct and collocate the system matrices into an appropriate set of simultaneous equations. In Chapter 6 of this text, a general computer code formulated from the information in Chapters 2, 3, and 4 is presented. The computer program uses the same assembly procedure demonstrated in this constant element example.

3.3 Quadratic Element

Quadratic elements are often used to achieve a more accurate representation of the geometry of the problem domain and provide a second order approximation of the function over each element.

3.3.1 Isoparametric Quadratic Elements

The boundary is defined by a series of n discrete nodal points. A general quadratic element will be defined by a continuous set of three nodal points as shown in Figure 3.4. A second order distribution of the boundary quantities and the geometry will be assumed over each element. The resulting quadratic element

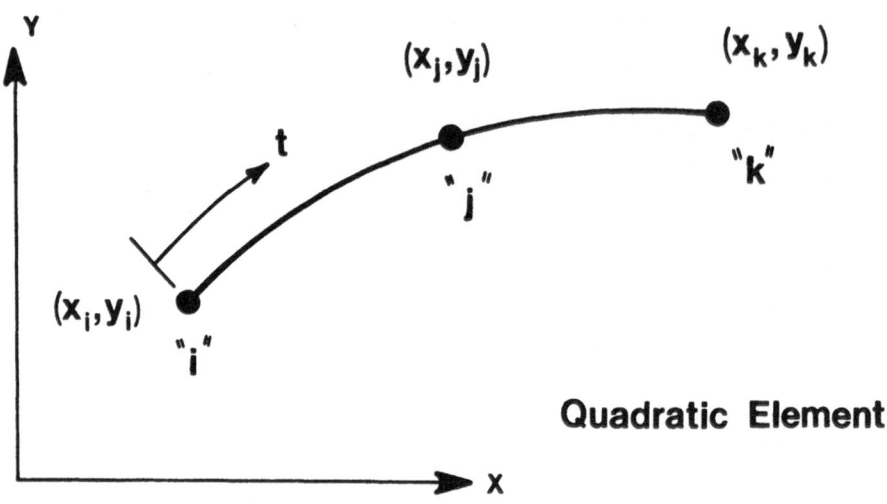

Quadratic Element

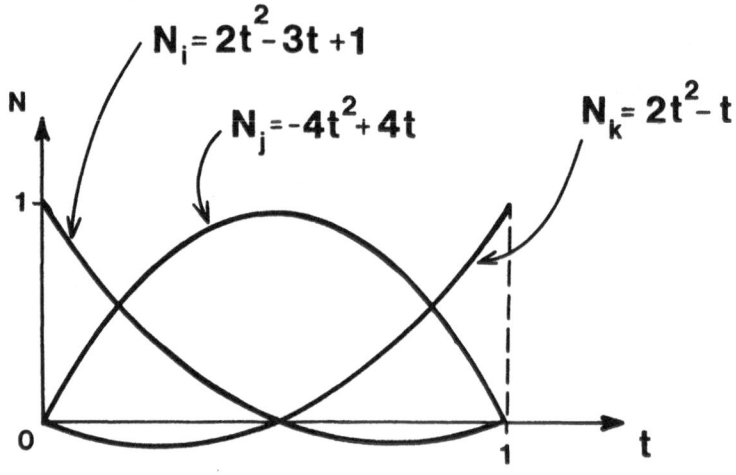

Shape Functions

Figure 3.4 Quadratic Element Nomenclature

is C_0 continuous in each of the four boundary quantities ψ, ω, ψ', and ω' between elements.

The quadratic element is transformed from two-dimensional Cartesian coordinates (x,y) into a single parameter curvilinear coordinate, t. The values of any of the boundary quantities ψ, ω, ψ', and ω' as well as the Cartesian coordinates of the approximate geometry are defined in terms of their discrete nodal values and a shape function set <N>. The shape functions for a quadratic element in terms of the parameter t are given as

$$N_i = 2t^2 - 3t + 1 \qquad N_j = -4t^2 + 4t \qquad N_k = 2t^2 - t \qquad (3.49)$$

Using the shape functions defined in Equation (3.49), the variation of the Cartesian coordinates x and y as a function of the parameter t may be written as

$$x(t) = N_i x_i + N_j x_j + N_k x_k$$

$$y(t) = N_i y_i + N_j y_j + N_k y_k \qquad (3.50)$$

The evaluation of the integrations in Equations (2.30) - (2.33) in the parameter space requires a transformation using a simple Jacobian defined as

$$|J| = \sqrt{\left(\frac{dx}{dt}\right)^2 + \left(\frac{dy}{dt}\right)^2} = \frac{dS}{dt} \qquad (3.51)$$

Unlike the linear element the Jacobian for a quadratic formulation is a function of the parameter t. An elemental form of the Jacobian may be derived from the discrete nodal coordinates and the shape functions. Equation (3.50) may be rewritten in the following form:

$$x(t) = A_x t^2 + B_x t + x_i$$

$$y(t) = A_y t^2 + B_y t + y_i \qquad (3.52)$$

where the element constants A_x, A_y, B_x, and B_y are defined as

$$A_x = 2x_i - 4x_j + 2x_k \qquad B_x = -3x_i + 4x_j - x_k$$

$$(3.53)$$

$$A_y = 2y_i - 4y_j + 2y_k \qquad B_y = -3y_i + 4y_j - y_k$$

The derivatives of the x and y with respect to t are calculated and substituted into the expression for the Jacobian. The resulting form of the Jacobian is

$$|J| = \left(4At^2 + 2Bt + C \right)^{1/2} \qquad (3.54)$$

where the constants A, B, and C are defined as

$$A = A_x^2 + A_y^2 \qquad B = 2(A_x B_x + A_y B_y) \qquad C = B_x^2 + B_y^2 \qquad (3.55)$$

The position vector $|\vec{q}_p - \vec{q}|$ is transformed into a function of the parameter t by the relationship defined in Equation (3.6). Substituting the quadratic form of x(t) and y(t) into Equation (3.6) results in

$$|\vec{q}_p - \vec{q}|^2 = At^4 + Bt^3 + (C+E)t^2 + Ft + D \equiv X \qquad (3.56)$$

where the constants D, E, and F are defined as

$$D = D_x^2 + D_y^2 \qquad E = 2(A_x D_x + A_y D_y) \qquad F = 2(B_x D_x + B_y D_y) \qquad (3.57)$$

$$D_x = x_i - x_p \qquad D_y = y_i - y_p$$

Substituting the relationships for the Jacobian, the position vector, and the quadratic shape functions into the integrals defined in Equations (2.30) - (2.33) results in the following expressions:

$$H_{pe} = \frac{1}{2\pi} \int_0^1 (N_i \psi_i + N_j \psi_j + N_k \psi_k)$$

$$\cdot \frac{(A_x t^2 + B_x t + D_x)n_x + (A_y t^2 + B_y t + D_y)n_y}{X} |J| \, dt \qquad (3.58)$$

$$G_{pe} = \frac{1}{4\pi} \int_0^1 (N_i \psi_i' + N_j \psi_j' + N_k \psi_k') \ln X \ |J| \ dt \qquad (3.59)$$

$$L_{pe} = \frac{1}{8\pi} \int_0^1 (N_i \omega_i + N_j \omega_j + N_k \omega_k)(\ln X - 1)$$

$$\cdot \ ((A_x t^2 + B_x t + D_x)n_x + (A_y t^2 + B_y t + D_y)n_y)|J|dt \qquad (3.60)$$

$$K_{pe} = \frac{1}{16\pi} \int_0^1 (N_i \omega_i' + N_j \omega_j' + N_k \omega_k') \ X \ (\ln X - 2) \ |J| \ dt \qquad (3.61)$$

The direction cosines n_x and n_y are functions of the parameter t, the Jacobian, and the element constants, and are defined as

$$n_x = \frac{2A_y t + B_y}{|J|} \qquad\qquad n_y = - \frac{2A_x t + B_x}{|J|} \qquad (3.62)$$

Equations (3.58) - (3.61) may be evaluated numerically for a general source point (x_p, y_p). However, a special form of Equation (3.59) is required when the source point is a member of the element being integrated since the natural logarithmic part of the fundamental solution is singular at that point. Since a quadratic element is located by three discrete nodal points, there are three locations where Equation (3.59) will become singular. If the source point (x_p, y_p) is equal to the nodal point (x_i, y_i), Equation (3.56) reduces to

$$X = At^4 + Bt^3 + Ct^2 \qquad (3.63)$$

Therefore Equation (3.59) may be rewritten as

$$G_{pe} = \frac{1}{4\pi} \int_0^1 <N>\{\psi'\} \ \ln(\ At^4 + Bt^3 + Ct^2 \ )|J|dt \qquad (3.64)$$

By factoring a t^2 from the logarithmic function the integral may be separated into a singular part and a non-singular part

$$G_{pe} = \frac{1}{4\pi} \int_0^1 <N>\{\psi'\} \ln\left(At^2 + Bt + C \right) |J| dt$$

$$+ \frac{1}{2\pi} \int_0^1 <N>\{\psi'\} \ln t \; |J| dt \qquad\qquad (3.65)$$

The second term in Equation (3.65) may be integrated using any appropriate logarithmic quadrature scheme (Stroud and Secrest, 1966).

Evaluating the singular integral for a quadratic element when the source point (x_p, y_p) is equal to the nodal point (x_j, y_j) requires a different approach. The position vector definition, Equation (3.56), cannot be factored as before. Instead, the shape functions will be rewritten as a function of a new parameter s, that varies from -1 to 1. This maneuver effectively divides the integration in half where each piece contains a term that becomes singular at s = 0. The transformation from t space to s space is given as

$$s = 2t - 1 \qquad\qquad (3.66)$$

The new shape functions in s space are

$$N_i^* = \frac{1}{2}(s^2 - s) \qquad N_j^* = 1 - s^2 \qquad N_k^* = \frac{1}{2}(s^2 + s) \qquad (3.67)$$

Substituting the change of variables for s into the position vector and the Jacobian results in

$$X^* = \frac{A}{16} s^4 + \left(\frac{A}{4} + \frac{B}{8}\right)s^3 + \left(\frac{3(A+B)}{8} + \frac{(C+E)}{4}\right)s^2 \qquad (3.68)$$

$$|J^*| = \left(A s^2 + (2A+B) s + A + B + C \right)^{1/2} \qquad (3.69)$$

Equation (3.59) may be rewritten in the new parameter s as:

$$G_{pe} = \frac{1}{8\pi} \int_0^1 <N^*>\{\psi'\} \ln X^* \; |J^*| \; ds$$

$$- \frac{1}{8\pi} \int_0^{-1} <N^*>\{\psi'\} \ln X^* \; |J^*| \; ds \qquad (3.70)$$

Another change of variable is required to transform the limits of the second integral of Equation (3.70). A new parameter u is defined by u = -s. The shape functions, the position vector, and the Jacobian are rewritten as functions of the parameter u, and defined as

$$N_i^{**} = \frac{1}{2}(u^2+u) \qquad N_j^{**} = 1-u^2 \qquad N_k^{**} = \frac{1}{2}(u^2-u) \qquad (3.71)$$

$$X^{**} = \frac{A}{16}u^4 - (\frac{A}{4} + \frac{B}{8})u^3 + (\frac{3(A+B)}{8} + \frac{(C+E)}{4})u^2 \qquad (3.72)$$

$$|J^{**}| = (Au^2 - (2A+B)u + A + B + C)^{1/2} \qquad (3.73)$$

Substituting Equations (3.71) - (3.73) into the second integral of Equation (3.70) yields the final form of the required integration:

$$G_{pe} = \frac{1}{8\pi} \int_0^1 <N^*>\{\psi'\} \ln X^* |J^*| \, ds$$

$$+ \frac{1}{8\pi} \int_0^1 <N^{**}>\{\psi'\} \ln X^{**} |J^{**}| \, du \qquad (3.74)$$

Both integrals of Equation (3.74) are of a form that has been confronted previously. Each may be decomposed into a singular part and nonsingular part similar to the procedure used in Equation (3.65) and evaluated with the appropriate numerical quadrature scheme.

The final case is when the source point (x_p, y_p) is equal to the nodal point (x_k, y_k). The logarithmic terms become singular at t = 1. Defining a change of variable, t = 1 - v, effectively reverses the limits of the integration. The shape functions, the position vector, and the Jacobian are redefined as

$$N_i^{***} = 2v^2-v \qquad N_j^{***} = -4v^2+4v \qquad N_k^{***} = 2v^2-3v+1 \qquad (3.75)$$

$$X^{***} = Av^4 - (4A+B)v^3 + (6A+3B+C+E)v^2 \qquad (3.76)$$

$$|J^{***}| = (4Av^2 + (2B+8A)v + 4A + 2B + C)^{1/2} \qquad (3.77)$$

The substitution of Equations (3.75) - (3.77) into Equation (3.59) allows the integrations to be rewritten as

$$G_{pe} = \frac{1}{8\pi} \int_0^1 <N^{***}>\{\psi'\} \ln X^{***} |J^{***}| \, dv \qquad (3.78)$$

By factoring a v^2 out of the logarithmic term in Equation (3.78), the integral may be written as two separate integrals, of which only one is singular. This procedure is identical to that used in the previous analysis when the source point is equal to (x_i, y_i) except the shape functions are reversed and the position vector and the Jacobian have slightly different forms.

Boundary element analysis using general quadratic elements presents some of the same problems as do their finite element counterparts. Irregular spacing of the element nodes can lead to the development of "overspill" whereby the distribution of the geometry is characterized by kinks and spurious wiggles (Zienkiewicz, 1977). Errors associated with the evaluation of the integrals over a general quadratic element by numerical quadrature are minimal. However, the calculation of internal points very near a quadratic element suffers from the same inconsistencies and inaccuracies as those associated with a general linear element.

3.3.2 Subparametric Quadratic Elements

A subparametric form of the quadratic element has certain tactical advantages in some problems over the isoparametric form. If the geometry of the problem domain is linear or a lower order approximation of the surface is assumed, the Jacobian becomes a constant and may be factored from the integrations of Equations (3.58) - (3.61). The resulting subparametric form may be evaluated analytically. The additional intrinsic advantages of an exact analysis may offset some of the disadvantages associated with a lower order approximation of the geometry.

The boundary is divided into n straight line segments where the distribution of each of the boundary quantities ψ, ω, ψ', and ω' is a function of three equally spaced discrete nodal values and the shape functions given in Equation (3.49). The resulting subparametric formulation is C_0 continuous between elements. Since the geometry is linear over each element, Equation (3.50) may be rewritten as

$$x(t) = (1-t)x_i + tx_k \qquad y(t) = (1-t)y_i + ty_k \qquad (3.79)$$

The Jacobian and the position vector defined in Equations (3.51) and (3.56) must be rewritten using the relationships in Equation (3.79). The resulting subparametric form of the Jacobian and the position vector is identical to that of a linear element, given in Equations (3.5) and (3.8), except for the element constants. The Jacobian and the position vector are defined as

$$|J| = \sqrt{A} = l_e \qquad l_e \equiv \text{Element Length} \qquad (3.80)$$

$$X = At^2 + Bt + C \qquad (3.81)$$

where the element constants A, B, and C are given by Equation (3.55) with the following corrections:

$$A_x = x_k - x_i \qquad A_y = y_k - y_i \qquad (3.82)$$

At this point, the procedure continues in the same way as the analytical analysis for the linear element. The linear shape functions in Equations (3.10) - (3.13) are replaced by their quadratic counterparts. The resulting integrands are of the form t^n, t^n/X, and $t^n \ln X$. Complementing the previous integration table defined in Equations (3.15) - (3.24) with the following additions will provide the necessary components for an exact analysis:

$$I_6 = \int_0^1 \frac{t^6}{X} dt = (\frac{1}{5} - BI_5 - CI_4)/A \qquad (3.83)$$

$$L_4 = \int_0^1 t^3 \ln X \, dt = \frac{1}{5} \left(\ln(A+B+C) - 2AI_6 - BI_5 \right) \tag{3.84}$$

The analytical expressions for the integration over a subparametric quadratic element of the integrals defined in Equations (2.30) - (2.33) are

$$H_{pe} = \frac{1}{2\pi} \frac{e}{} \left(H_i \psi_i + H_j \psi_j + H_k \psi_k \right) \tag{3.85}$$

$$G_{pe} = \frac{1}{4\pi} \frac{e}{} \left(G_i \psi_i' + G_j \psi_j' + G_k \psi_k' \right) \tag{3.86}$$

$$L_{pe} = \frac{1}{8\pi} \frac{e}{} \left(L_i \omega_i + L_j \omega_j + L_k \omega_k \right) \tag{3.87}$$

$$K_{pe} = \frac{1}{16\pi} \frac{e}{} \left(K_i \omega_i' + K_j \omega_j' + K_k \omega_k' \right) \tag{3.88}$$

where

$$H_i = E(2I_2 - 3I_1 + I_0)$$
$$H_j = 4E(-I_2 + I_1) \tag{3.89}$$
$$H_k = E(2I_2 - I_1)$$

$$G_i = 2L_2 - 3L_1 + L_0$$
$$G_j = 4(-L_2 + L_1) \tag{3.90}$$
$$G_k = 2L_2 - L_1$$

$$L_i = E(2L_2 - 3L_1 + L_0 - \frac{1}{6})$$
$$L_j = 4E(-L_2 + L_1 - \frac{1}{6}) \tag{3.91}$$
$$L_k = E(2L_2 - L_1 - \frac{1}{6})$$

$$K_i = 2A\left(L_4 - \frac{2}{5}\right) + (2B-3A)\left(L_3 - \frac{1}{2}\right) + (A-3B+2C)\left(L_2 - \frac{2}{3}\right)$$

$$+ (B-3C)(L_1 - 1) + C(L_0 - 2)$$

$$K_j = -A\left(L_4 - \frac{2}{5}\right) + (A-B)\left(L_3 - \frac{1}{2}\right) + (B-C)\left(L_2 - \frac{2}{3}\right)$$

$$+ C(L_1 - 1)$$

$$K_k = 2A\left(L_4 - \frac{2}{5}\right) + (2B-A)\left(L_3 - \frac{1}{2}\right) + (2C-B)\left(L_2 - \frac{2}{3}\right)$$

$$- C(L_1 - 1)$$

(3.92)

Unlike the isoparametric form, the analytical subparametric formulation explicitly handles the logarithmic singularity when the source point is at an end node. However, if the source point (x_p, y_p) coincides with the middle node (x_j, y_j), the shape functions must be transformed like their isoparametric counterparts. The element is divided in two about the (x_j, y_j) point. The transformed shape functions for each segment are previously defined in Equations (3.67) and (3.71). The form of the analytical expression for this case is the same as that defined in Equations (3.85) - (3.88). However, the integration table must be recompiled with the change of variable, the result being a different form of the integration components given in Equations (3.89) - (3.92). The element constants are redefined for this particular case in the following form:

$$A_1 = \frac{A}{16} \qquad B_1 = \frac{A}{4} + \frac{B}{8} \qquad C_1 = \frac{3(A+C)}{8} + \frac{B+C}{4} \tag{3.93}$$

$$A_2 = A_1 \qquad B_2 = -B_1 \qquad C_2 = C_1 \tag{3.94}$$

The integration table may now be recompiled with the new form of the element constants and is listed below: $\qquad (i = 1,2)$

$$I_{01} = \frac{2\gamma}{(4A_1C_1 - B_1^2)^{1/2}} \qquad I_{02} = \frac{2\gamma}{(4A_2C_2 - B_2^2)^{1/2}} \tag{3.95}$$

$$I_{1i} = (\ln(A_i + B_i + C_i) - \ln C_i - B_i I_{0i})/2A_i \tag{3.96}$$

$$I_{ni} = \frac{1}{n-1} - B_i I_{n-1,i} - C_i I_{n-2,i} \tag{3.97}$$

$$L_{ni} = \frac{1}{n+1} \left(\ln(A_i + B_i + C_i) - 2A_i I_{n+2,i} - B_i I_{n+1,i} \right) \tag{3.98}$$

The original set of integration components, Equations (3.89) - (3.92), are redefined as

$$H_i = \frac{E}{2} \left(I_{21} - I_{11} + I_{22} + I_{12} \right)$$

$$H_j = E \left(I_{01} - I_{21} + I_{02} - I_{22} \right) \tag{3.99}$$

$$H_k = \frac{E}{2} \left(I_{21} + I_{11} + I_{22} - I_{12} \right)$$

$$G_i = \frac{1}{2} \left(L_{21} - L_{11} + L_{22} + L_{12} \right)$$

$$G_j = L_{01} - L_{21} + L_{02} - L_{22} \tag{3.100}$$

$$G_k = \frac{1}{2} \left(L_{21} + L_{11} + L_{22} - L_{12} \right)$$

$$L_i = \frac{E}{2} \left(L_{21} - L_{11} + L_{22} + L_{12} - \frac{2}{3} \right)$$

$$L_j = E \left(L_{01} - L_{21} + L_{02} - L_{22} - \frac{4}{3} \right) \tag{3.101}$$

$$L_k = \frac{E}{2} \left(L_{21} + L_{11} + L_{22} - L_{12} - \frac{2}{3} \right)$$

$$K_i = \frac{1}{2} \left(A_1 (L_{41} + L_{42} - \frac{4}{5}) + (B_1 - A_1)(L_{31} - L_{32}) \right.$$
$$\left. (C_1 - B_1)(L_{21} + L_{22} - \frac{4}{3}) - C_1(L_{11} - L_{12}) \right)$$

$$K_j = -A_1 (L_{41} + L_{42} - \frac{4}{5}) - B_1(L_{31} - L_{32}) + (A_1 - C_1)$$
$$\cdot (L_{21} + L_{22} - \frac{4}{3}) + B_1(L_{11} - L_{12}) \tag{3.102}$$
$$+ C_1(L_{01} + L_{02} - 4)$$

$$K_k = \frac{1}{2} \left(A_1(L_{41} + L_{42} - \frac{4}{5}) + (B_1 + A_1)(L_{31} - L_{32}) \right.$$
$$\left. + (C_1 + B_1)(L_{21} + L_{22} - \frac{4}{3}) + C_1(L_{11} - L_{12})) \right.$$

The analytical expressions for the integrations over a quadratic subparametric element presented, although lengthy, reduce the time required to formulate system matrices as compared to the isoparametric version. Accuracy of the calculation of the field variables at internal points very close to boundary is also improved. This type of element is very useful when the geometry of the problem is composed of linear segments. In this particular case, the subparametric quadratic element is of the same order as an isoparametric element.

3.4 Overhauser Elements

Linear and quadratic elements are generally sufficiently accurate to describe many engineering problems. A variety of curved geometries are well represented by a standard quadratic element or a combination of quadratic and linear elements. A common drawback to both types of formulations is the lack of derivative continuity between elements. Several different types of spline elements that are C_1 continuous have been used for various purposes (Kreyszig, 1983).

Cubic splines provide derivative continuity, but are computationally inefficient and cumbersome (Ligget and Salmon, 1981). Most types of formulations require an additional variable at each end node which enforces the prescribed derivative continuity.

Overhauser (1968) introduced a cubic parametric representation of a curve by blending two parametric quadratic curves. Derivative continuity is implicitly defined by the curve. In this section, a formulation of the Overhauser curve developed by Brewer (1977; Brewer and Anderson, 1977), as shown in Figure 3.5, will be implemented. The parametric curve, c(t), is a blend of two quadratic curves, p(r) and q(s), where t, r, and s are curvilinear parameters along their respective curves. Note that the Overhauser curve is defined between "regular" points j and k. The "extra" nodes i and l are used to maintain the derivative continuity.

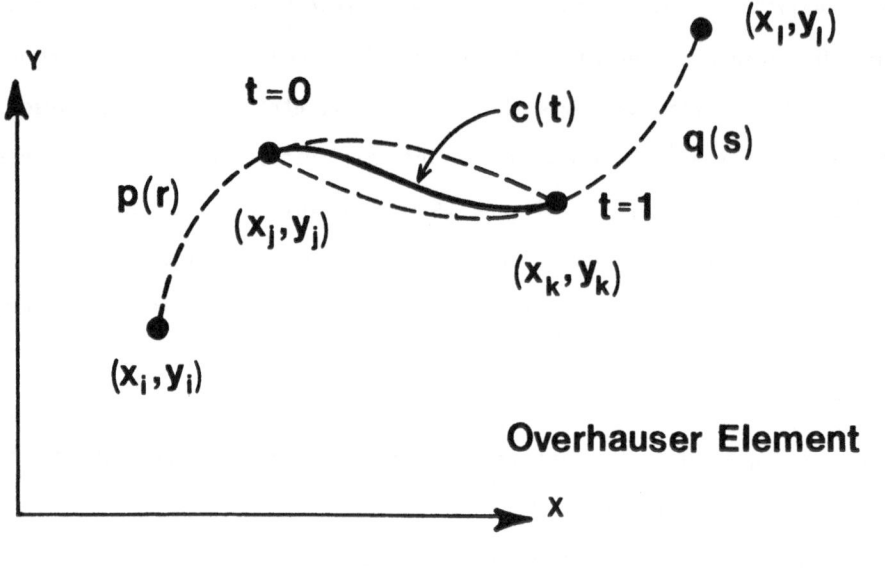

Overhauser Element

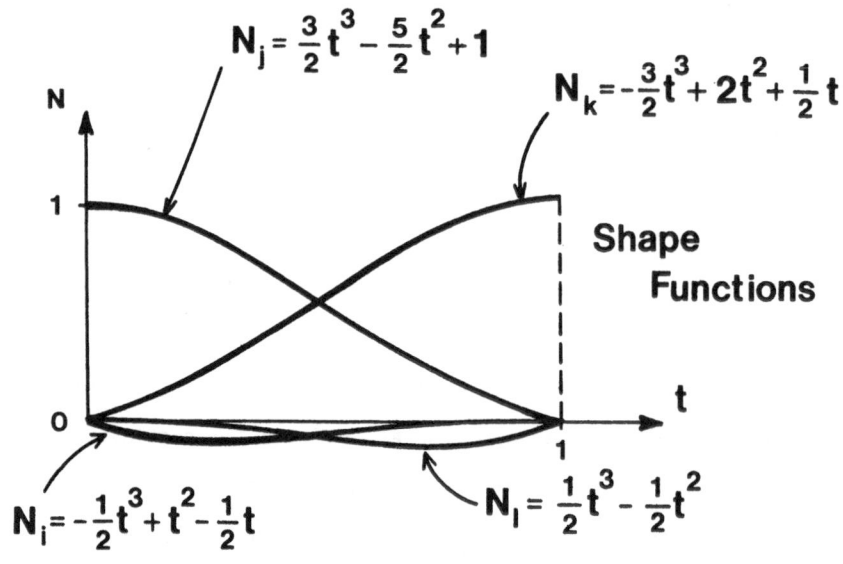

Shape Functions

Figure 3.5 Overhauser Element Nomenclature

3.4.1 Isoparametric Overhauser Element

The boundary is divided into a series of n discrete nodal points, a procedure similar to that used to model the boundary by quadratic elements, except with four associated nodes. The value of any of the boundary quantities ψ, ω, ψ', ω', as well as the Cartesian coordinates of the approximate geometry, are defined in terms of four consecutive discrete nodal values (two of which do not reside on the element in question) and the shape functions. In terms of the parameter t, the shape functions are given as

$$N_i = -\frac{1}{2}t^3 + t^2 - \frac{1}{2}t \qquad N_j = \frac{3}{2}t^3 - \frac{5}{2}t^2 + 1$$

$$N_k = -\frac{3}{2}t^3 + 2t^2 + \frac{1}{2}t \qquad N_l = \frac{1}{2}t^3 - \frac{1}{2}t^2$$

(3.103)

Using these shape functions, the Cartesian coordinates x and y as a function of the parameter t are written as

$$x(t) = N_i x_i + N_j x_j + N_k x_k + N_l x_l$$

$$y(t) = N_i y_i + N_j y_j + N_k y_k + N_l y_l$$

(3.104)

The development of the Overhauser element follows the same procedure as the quadratic element formulation. Therefore, the necessary form of the position vector and the Jacobian operator are defined as

$$|J| = \left(9At^4 + 6A_b t^3 + (3A_c + 4B)t^2 + 2B_c + C \right)^{1/2}$$

(3.105)

$$X = At^6 + A_b t^5 + (A_c + B)t^4 + (A_c + A_d)t^3$$

$$+ (B_c + C)t^2 + C_d t + D$$

(3.106)

where the element constants A, B, C, D, A_b, A_c, A_d, B_c, B_d, and C_d are given as

$$A = A_x^2 + A_y^2 \qquad B = B_x^2 + B_y^2 \qquad C = C_x^2 + C_y^2$$

$$A_b = 2(A_x B_x + A_y B_y) \qquad A_c = 2(A_x C_x + A_y C_y)$$

$$A_d = 2(A_x D_x + A_y D_y) \qquad B_c = 2(B_x C_x + B_y C_y)$$

$$B_d = 2(B_x D_x + B_y D_y) \qquad C_d = 2(C_x D_x + C_y D_y)$$

$$A_x = -\frac{1}{2} x_i + \frac{3}{2} x_j - \frac{3}{2} x_k + \frac{1}{2} x_l \qquad (3.107)$$

$$B_x = x_i - \frac{5}{2} x_j + 2 x_k - \frac{1}{2} x_l$$

$$C_x = -\frac{1}{2} x_i + \frac{1}{2} x_k \qquad D_x = x_j - x_p$$

$$A_y = -\frac{1}{2} y_i + \frac{3}{2} y_j - \frac{3}{2} y_k + \frac{1}{2} y_l$$

$$B_y = y_i - \frac{5}{2} y_j + 2 y_k - \frac{1}{2} y_l$$

$$C_y = -\frac{1}{2} y_i + \frac{1}{2} y_k \qquad D_y = y_j - y_p$$

Substituting the Overhauser shape functions, the Jacobian, and position vector relationships of Equations (3.105) and (3.106) into Equations (2.30) - (2.33) results in the following expressions for the required integrations:

$$H_{pe} = \frac{1}{2\pi} \int_0^1 (N_i \psi_i + N_j \psi_j + N_k \psi_k + N_l \psi_l)$$
$$\cdot \frac{X_p n_x + Y_p n_y}{X} |J| \, dt \qquad (3.108)$$

$$G_{pe} = \frac{1}{4\pi} \int_0^1 (N_i \psi_i' + N_j \psi_j' + N_k \psi_k' + N_l \psi_l') \ln X \, |J| \, dt \qquad (3.109)$$

$$L_{pe} = \frac{1}{8\pi} \int_0^1 (N_i \omega_i + N_j \omega_j + N_k \omega_k + N_l \omega_l)$$
$$\cdot (X_p n_x + Y_p n_y) |J| \, dt \qquad (3.110)$$

$$K_{pe} = \frac{1}{16\pi} \int_0^1 (N_i\omega_i' + N_j\omega_j' + N_k\omega_k' + N_l\omega_l')$$

$$\cdot X (\ln X - 2) |J| \, dt \tag{3.111}$$

where X_p, Y_p, and the direction cosine are functions of the parameter t, the Jacobian, and the element constants:

$$X_p = A_x t^3 + B_x t^2 + C_x t + D_x$$

$$Y_p = A_y t^3 + B_y t^2 + C_y t + D_y \tag{3.112}$$

$$n_x = \frac{3A_y t^2 + 2B_y t + C_y}{|J|}$$

$$n_y = - \frac{3A_x t^2 + 2B_x t + C_x}{|J|}$$

Equations (3.108) - (3.111) may be evaluated numerically for a general source point (x_p, y_p). However, a special form of Equation (3.93) is necessary when the source point is equal to either point (x_j, y_j) or (x_k, y_k). When this situation occurs, the natural logarithmic terms of the fundamental solution become singular, a circumstance reminiscent of the preceding linear and quadratic formulations. If the source point (x_p, y_p) is equal to the nodal point (x_j, y_j) the expression for the position vector, Equation (3.106), reduces to

$$X = At^6 + A_b t^5 + (A_c + B)t^4 + B_c t^3 + Ct^2 \tag{3.113}$$

By factoring a t^2 from the logarithmic function the integral in Equation (3.109) is separated into a singular part and a nonsingular part:

$$G_{pe} = \frac{1}{4\pi} \int_0^1 <N>\{\psi'\} \ln(At^4 + A_b t^3 + (A_c + B)t^2$$

$$+ B_c t + C) |J| \, dt + \frac{1}{2\pi} \int_0^1 <N>\{\psi'\} \ln t \, |J| \, dt \tag{3.114}$$

The second integral of Equation (3.114) has effectively isolated the singularity problem, and the integral may be evaluated using any logarithmic quadrature scheme. Another singularity occurs in Equation (3.109) at $t = 1$, when the source point (x_p, y_p) is equal to the nodal point (x_k, y_k). A change of variables, $t = 1 - u$, will reverse the limits of the integration such that the singularity occurs at $u = 0$. The shape functions, the position vector, and the Jacobian may be redefined in terms of the parameter u as

$$N_i^* = \frac{1}{2}\left(s^3 - s^2 \right) \qquad N_j^* = -\frac{3}{2}s^3 + 2s^2 + \frac{1}{2}s$$

$$N_k^* = \frac{3}{2}s^3 - \frac{5}{2}s^2 + 1 \qquad N_l^* = -\frac{1}{2}s^3 + s^2 - \frac{1}{2}s \tag{3.115}$$

$$X^* = A s^6 + (A_b - 6A)s^5 + (15A + A_b + A_c + B)s^4$$
$$+ (-20A - 10A_b - 4(A_c + B) - A_d + B_c)s^3$$
$$+ (15A + 10A_b + 6(A_c + B) + 3(A_d + B_c) + B_c + C)s^2 \tag{3.116}$$

$$|J^*| = (9A\,s^4 - (36A + 6A_b)s^3 + (54A + 4B + 3A_c + 18A_c)s^2$$
$$- (36A + 8B + 6A_c + 2B_c + 18A_d)s$$
$$+ (9A + 4B + C + 3A_c + 2B_c + 6A_b))^{\frac{1}{2}} \tag{3.117}$$

By substituting Equations (3.99) - (3.100) into Equation (3.93) the integration is rewritten in the following form:

$$G_{ps} = \frac{1}{4\pi} \int_0^1 <N^*>\{\psi'\}\, \ln X^*\, |J^*|\, ds \tag{3.118}$$

By factoring the s^2 out of the logarithmic term of Equation (3.118) the integral may be separated into a singular part and a nonsingular part. The result is similar to what happened in Equation (3.114), except that the shape functions are reversed, and the Jacobian and the position vector have different forms. As before, the singular part of the integration may be evaluated with any appropriate logarithmic quadrature scheme.

Boundary element analysis using Overhauser elements provides many interesting advantages over other spline elements or lower order elements. The approximation to the distribution of the field variables and the geometry over an element is represented by cubic order shape functions which are C_1 continuous between elements. The computational implementation of an Overhauser element is much simpler than a standard quadratic or a cubic type element. The characteristic of "overspill" evident in higher order elements is far less pronounced in the Overhauser formulation. This is demonstrated in Figure 3.6, where several cases of abnormal node spacing are presented. However, it is advisable to avoid abrupt changes in the noding spacing.

The main disadvantage of the Overhauser element occurs when modeling discontinuities in the geometry such as corners. Since the element is designed to model C_1 continuous surfaces, it does not accurately represent abrupt changes in the geometry. Therefore, a special form of the Overhauser element is used for modeling corners. This is accomplished by "double noding" where one of the "extra" nodes is defined to be at the same location as one of the regular nodes. Equally accurate results are obtained when the Overhauser is linked to either a quadratic or a linear element to describe a corner.

3.4.2 Subparameteric Overhauser Elements

The subparametric form of the Overhauser element has the same advantages as its quadratic counterpart. If the problem under consideration is segmentally linear or the geometry is assumed linear, the Jacobian operator becomes a constant and can be factored from the required integrations. The resulting integrals may be evaluated analytically. Therefore, it is possible to have the intrinsic advantage of an exact analysis with a cubic order C_1 continuous approximation for the boundary quantities.

● **Nodes on the element**

○ **Nodes that define the derivatives (not on element)**

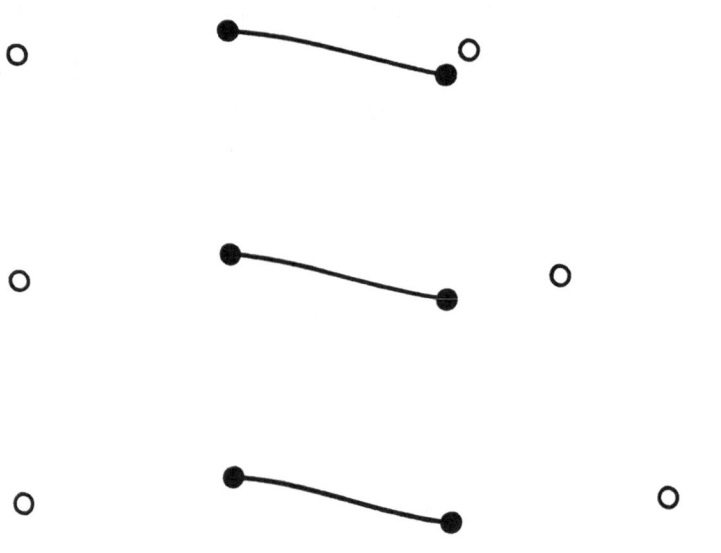

Figure 3.6 Behavior of Overhauser Element with Abnormal Node Spacing

The boundary is divided into n straight line segments. The subparametric form of the Overhauser element will be defined by four consecutive equally spaced nodes. Since the geometry is linear, the variation of the Cartesian coordinates with the parameter t given in Equation (3.88) is redefined as

$$x(t) = (1-t)x_j + tx_k \qquad y(t) = (1-t)y_j + ty_k \qquad (3.119)$$

The Jacobian and the position vector defined in Equations (3.105) and (3.106) are recalculated using the relationship in Equation (3.119). The resulting subparametric form of the Jacobian and the position vector are similar to those defined for a subparametric quadratic element, given in Equations (3.80) and (3.81), except that the element constants are different. The Jacobian and the position vector are

$$|J| = \sqrt{A} = l_e \qquad l_e \equiv \text{Element Length} \qquad (3.120)$$

$$X = At^2 + Bt + C \qquad (3.121)$$

where the element constants A, B, and C are given in Equation (3.54) with the following corrections

$$A_x = x_k - x_j \qquad\qquad A_y = y_k - y_j$$
$$\qquad\qquad\qquad\qquad\qquad\qquad\qquad (3.122)$$
$$B_x = x_j - x_p \qquad\qquad B_y = y_j - y_p$$

At this point, the procedure continues in an identical manner to that of the linear element. The linear shape function in Equations (3.10) - (3.13) are replaced by the Overhauser shape functions given in Equation (3.108). The resulting integral contains terms of the form t^n, t^n/X, and $t^n \ln X$, identical to those found in the linear element, except to a higher degree. Therefore, the integration table defined in Equations (3.15) - (3.24) is supplemented with the following additions to provide a complete set of components for an exact analysis:

$$I_7 = \int_0^1 \frac{t^7}{X} dt = (\frac{1}{6} - BI_6 - CI_5)/A \qquad (3.123)$$

$$L_5 = \int_0^1 t^5 \ln X \, dt = \frac{1}{6} (\ln(A+B+C) - 2AI_7 - BI_6) \qquad (3.124)$$

The analytical expression for the integrations defined in Equations (2.30) – (2.33) using a subparametric Overhauser element are

$$H_{pe} = \frac{1_e}{2\pi} (H_i \psi_i + H_j \psi_j + H_k \psi_k + H_1 \psi_1) \qquad (3.125)$$

$$G_{pe} = \frac{1_e}{4\pi} (G_i \psi_i' + G_j \psi_j' + G_k \psi_k' + G_1 \psi_1') \qquad (3.126)$$

$$L_{pe} = \frac{1_e}{8\pi} (L_i \omega_i + L_j \omega_j + L_k \omega_k + L_1 \omega_1) \qquad (3.127)$$

$$K_{pe} = \frac{1_e}{16\pi} (K_i \omega_i' + K_j \omega_j' + K_k \omega_k' + K_1 \omega_1') \qquad (3.128)$$

where

$$H_i = \frac{E}{2} (-I_3 + 2I_2 - I_1)$$

$$H_j = \frac{E}{2} (3I_3 - 5I_2 + 2I_0)$$

$$H_k = \frac{E}{2} (-3I_3 + 4I_2 + I_1) \qquad (3.129)$$

$$H_1 = \frac{E}{2} (I_3 - I_2)$$

$$G_i = \frac{1}{2} (- L_3 + 2L_2 - L_1)$$

$$G_j = \frac{1}{2} (3L_3 - 5L_2 + 2L_0)$$

$$G_k = \frac{1}{2} (- 3L_3 + 4L_2 + L_1) \qquad (3.130)$$

$$G_1 = \frac{1}{2} (L_3 - L_2)$$

$$L_i = \frac{E}{2} \left(-L_3 + 2L_2 - L_1 + \frac{1}{12} \right)$$

$$L_j = \frac{E}{2} \left(3L_3 - 5L_2 + 2L_0 - \frac{13}{12} \right)$$

$$L_k = \frac{E}{2} \left(-3L_3 + 4L_2 + L_1 - \frac{13}{12} \right)$$ \hspace{2cm} (3.131)

$$L_1 = \frac{E}{2} \left(L_3 - L_2 + \frac{1}{12} \right)$$

$$K_i = \frac{1}{2} \left(-A\left(L_5 - \frac{1}{3} \right) + (2A-B)\left(L_4 - \frac{2}{5} \right) + (2B-A-C)\left(L_3 - \frac{1}{2} \right) \right.$$

$$+ (2C-B)\left(L_2 - \frac{2}{3} \right) - C\left(L_1 - 1 \right))$$

$$K_j = \frac{1}{2} \left(3A\left(L_5 - \frac{1}{3} \right) + (3B-5A)\left(L_4 - \frac{2}{5} \right) + (3C-5B)\left(L_3 - \frac{1}{2} \right) \right.$$

$$+ (2A-5C)\left(L_2 - \frac{2}{3} \right) + 2B\left(L_1 - 1 \right) + 2C\left(L_0 - 2 \right))$$ \hspace{1cm} (3.132)

$$K_k = \frac{1}{2} \left(-3A\left(L_5 - \frac{1}{3} \right) + (4A-3B)\left(L_4 - \frac{2}{5} \right) + (A+4B-3C) \right.$$

$$\cdot \left(L_3 - \frac{1}{2} \right) + (B+4C)\left(L_2 - \frac{2}{3} \right) + C\left(L_1 - 1 \right))$$

$$K_1 = \frac{1}{2} \left(A\left(L_5 - \frac{1}{3} \right) + (B-A)\left(L_4 - \frac{2}{3} \right) + (C-B)\left(L_3 - \frac{1}{2} \right) \right.$$

$$- C\left(L_2 - \frac{2}{3} \right))$$

The analytical expressions for the integrations over the subparametric Over-hauser element presented significantly reduce the time required to formulate system matrices as compared to the isoparametric version. As with all analytical formulations presented, the calculation of the field variables at internal points very near the boundary are consistent and accurate. Since the Overhauser curve is a cubic, it often requires far fewer discrete nodes to effectively model the geometry and provide accurate reliable solutions. The analytical form of the element suffers the same restriction at abrupt changes in the geometry as the isoparametric version.

3.5 Modeling Corners

There are, of course, surface segments and boundary conditions that are not first derivative continuous and thus should not be splined. Modelling common

geometrical artifacts such as right-angle corners, where, say, derivative specified boundary conditions and potential-specified boundary conditions meet on adjacent surfaces, has long been a sore point in numerical modelling generally. In
analytical solutions, these points are referred to as "ill-behaved" locations.
In a numerical solution, some compromise must be effected in order to rationally
handle the contingency. Luckily, in boundary elements, there are three empirical
means for handling this problem.

The first technique is to use a discontinuous element on either side of the
offending point. Such an element is typified by the constant element presented
earlier. Since the element's node does not physically reside at the location of
the abrupt change of the boundary condition, the ambiguity does not explicitly
enter into the problem definition. The constant elements contact at the
offending point, and the nodal values, on either side of that point are given the
appropriate prescribed values. Thus, the problem of modeling the troublesome
point explicitly is empirically "bypassed."

Another method, suggested by Brebbia (1978), is to place two nodes of a
continuous element "very close" to each other on either side of the ill-defined
point. These nodes can be connected by an extremely small continuous element
several orders of magnitude shorter than any other element in the mesh; the
incompatible boundary conditions are prescribed at the nodes bounding this element. Again, the problem point is effectively left to do as it wishes in the
ensuing solution. Brebbia (1978) shows that such a strategy can provide quite
accurate results.

The third method, and the one which has proved most successful during the
course of this research, is to actually "double node" the corner. That is, allow
the end nodes of adjacent elements to actually reside physically at the same
point, but not have them physically connected. The two nodes will each have one
of the two different boundary conditions, depending upon the surface upon which
each is associated. Care must be taken in the integration strategy and the com-

puter program to handle such a case, but the results in all cases run by the authors have been outstanding. This method can actually be considered a compromise between the two preceding compromises mentioned.

A special problem arises when the Overhauser splining elements are used. Since the definition of the Overhauser requires nodes on adjacent elements to which it is smoothly joined, it is not appropriate for modelling the region adjacent to a corner. The strategy to be used in this case is to connect the Overhauser element with a non-splining element, the linear or the quadratic preferably, and allow the lower order element to model the corner in the third mentioned above. Excellent results have been obtained in this work using a mesh which is predominantly Overhauser elements, but with quadratic elements in the corners.

It should be mentioned in concluding this section that every strategy suggested here for handling "ill-behaved" points is empirical in nature, and results obtained using these methods should be scrutinized thoroughly before being accepted as final. Successful use of these techniques is clearly a function of experience and engineering judgment.

3.6 Summary and Concluding Remarks

Three general element types have been presented as possible representations for both the distribution of the field variables over the surface and the approximation of the boundary geometry. Analytical expressions for the required integrations over a linear element were given as well as the exact form for a special case of the quadratic and Overhauser elements.

Exact analytical analysis provides improved accuracy at internal point calculations, especially at points very near the boundary. The time required to compile the system matrices is greatly reduced while a high degree of accuracy is maintained in the solution. The subparametric assumption for the quadratic and Overhauser elements, which may reduce the order of the approximation for the geometry, is compensated for by the increased accuracy of an exact formulation. In Chapter 5, numerical examples will be presented comparing the three types of elements for both linear and curved geometries.

CHAPTER 4

DOMAIN DISCRETIZATION AND INTERNAL POINT CALCULATIONS

4.1 Introduction

The nonhomogeneous form of Equations (2.28) and (2.29) contains domain integrations as well as the surface integrals. Several ways of evaluating the domain
integration which will avoid any type of explicit domain discretization will be
developed in this chapter. This effort is an attempt to maintain the purity of
the boundary element formulation in the sense that only surface geometries need
be modeled.

Calculation of the values of the field variables ψ and ω at internal points
requires the evaluation of Equations (2.28) and (2.29). Analytical expressions
for the surface integrals in these equations have been presented in Chapter
III. Values of derivatives of the field variables ψ and ω at points in the
domain require the evaluation of a different set of integrals defined in
Equations (2.17) - (2.20) and (2.23) - (2.26). An exact analysis for a linear
isoparametric element will be developed. Both a general numerical form and an
analytical subparametric version of the quadratic and the Overhauser elements are
given.

4.2 Domain Discretization

Two methods of evaluating the most general nonhomogeneous terms defined in
Equations (2.34) and (2.35) are presented. The first technique uses a series of
Green's identities to transform special forms of the integration of the function
$f(x,y)$ over the domain V to a set of surface integrations. The second, more general technique uses a "domain fanning" quadrature scheme that does not require

implicit volume discretization. We begin the discussion with the special case of concentrated source terms.

4.2.1 Concentrated Source Terms

The simplest type of domain integral that can be handled without resorting to an explicit quadrature scheme is the case involving discrete source points. Physically, such domain terms might represent point loads applied to plates, or point sources/sinks in fluid flow problems. The mathematical representation of these physical idealizations via the Dirac delta renders the domain integrations trivial.

Consider the domain to be filled with n such point sources of strengths C_j located at points (x_j, y_j). The nonhomogeneous function may then be represented by

$$f(x,y) = \sum_{j=1}^{n} C_j \delta(\vec{p} - \vec{p}_j) \qquad (4.1)$$

where $\vec{p}_j$ locates the point (x_j, y_j). The domain integrals terms represented by Equations (2.34) and (2.35) are calculated quickly. Equation (2.34) becomes:

$$B1_i = \frac{1}{8\pi} \int_V \sum_{j=1}^{n} C_j \delta(\vec{p} - \vec{p}_j)(|\vec{q}_i - \vec{q}_j|^2 [\ln|\vec{q}_i - \vec{q}_j| - 1]) \, dV$$

$$= \frac{1}{8\pi} \sum_{j=1}^{n} C_j |q_i - q_j|^2 (\ln|q_i - q_j| - 1) \qquad (4.2)$$

Similarly, Equation (2.35) gives

$$B2_i = \frac{1}{2\pi} \int_V \sum_{j=1}^{n} C_j \delta(\vec{p} - \vec{p}_j) \ln|\vec{q}_i - \vec{q}_j| \, dV$$

$$= \frac{1}{2\pi} \sum_{j=1}^{n} C_j \ln|\vec{q}_i - \vec{q}_j| \qquad (4.3)$$

Therefore, the handling of point source terms is as simple as specifying their strengths and locations. The following example illustrates the synthesis and solution of a problem containing source terms.

4.2.1.1 Point Source Example - Constant Elements

Consider an extension of the creeping flow example presented in section 3.2.3 to that of one with a point source located at the center of the domain shown in Figure 3.3(a). The resulting equation governing the flow field in terms of the streamfunction ψ is given as:

$$\nabla^4\psi = \delta(y-1/2)\delta(x-1/2) \qquad (4.4)$$

As with the previous example in section 3.2.3, a very simple constant element formulation will be used in an attempt to demonstrate the assemble procedure necessary to form Equation (2.39). The details of constructing and collocating the system matrices are identical to those presented in section 3.2.3 and will not be repeated here. The effect of the point source is incorporated into the governing equation (2.39) through the terms {B1} and {B2} defined in Equations (2.34) and (2.35) respectively. As derived in section 4.2.1, the domain integrations defined in Equations (2.34) and (2.35) reduce to the analytical expressions given in Equations (4.2) and (4.3). By combining the point source terms calculated using Equations (4.2) and right hand side of the equations defined in Equation (3.47) and factoring out a $-1/2\pi$, a set of simultaneous equations approximating Equation (4.4) may be formed:

$$
\begin{bmatrix}
1.693 & 1.107 & -0.231 & 1.107 & 0.042 & -0.209 & 0.260 & -0.209 \\
0.335 & -3.142 & -0.209 & 0.927 & 0.176 & 0.000 & 0.176 & -0.231 \\
-0.039 & 1.107 & 0.000 & 1.107 & 0.260 & -0.209 & 0.042 & -0.209 \\
0.335 & 0.927 & -0.209 & -3.142 & 0.176 & -0.231 & 0.176 & 0.000 \\
0.000 & 0.000 & 0.927 & 0.000 & 1.693 & 1.107 & -0.039 & 1.107 \\
0.000 & 0.000 & 1.107 & 0.000 & 0.335 & -3.142 & 0.335 & 0.927 \\
0.000 & 0.000 & -3.142 & 0.000 & -0.039 & 1.107 & 1.693 & 1.107 \\
0.000 & 0.000 & 1.107 & 0.000 & 0.335 & 0.927 & 0.335 & -3.142
\end{bmatrix}
\begin{Bmatrix}
\psi'_1 \\
\psi_2 \\
\omega_3 \\
\psi_4 \\
\omega'_1 \\
\omega_2 \\
\omega'_3 \\
\omega_4
\end{Bmatrix}
=
$$

$$
= \left\{ \begin{array}{c} -0.92730 \\ -1.10715 \\ 3.14159 \\ -1.10715 \\ 0.00000 \\ 0.00000 \\ 0.00000 \\ 0.00000 \end{array} \right\} + \left\{ \begin{array}{c} 0.10582 \\ 0.10582 \\ 0.10582 \\ 0.10582 \\ 0.69315 \\ 0.69315 \\ 0.69315 \\ 0.69315 \end{array} \right\} \qquad (4.5)
$$

A comparison of the results for the unknown boundary quantities obtained from Equation (4.5) and the exact solution follows:

BEM   Exact

$$
\left\{ \begin{array}{c} \psi'_1 \\ \psi_2 \\ \omega_3 \\ \psi_4 \\ \omega'_1 \\ \omega_2 \\ \omega'_3 \\ \omega_4 \end{array} \right\} = \left\{ \begin{array}{c} -1.872 \\ 0.709 \\ -2.949 \\ 0.709 \\ 4.060 \\ -1.607 \\ -2.867 \\ -1.607 \end{array} \right\} \qquad \left\{ \begin{array}{c} -1.531 \\ 0.696 \\ -2.812 \\ 0.696 \\ 3.312 \\ -1.656 \\ -2.812 \\ -1.656 \end{array} \right\} \qquad (4.6)
$$

As before the discrepancy in the results is due to the crude mesh used for illustration purposes. The results of this example analysis demonstrated the relative ease of calculating and incorporating the effects of a point source into the procedure used to construct of the system assemble matrices. The purpose of this example was to illustrate these procedural concepts in a finite amount of space.

Neither the model used nor the simple constant element formulation are advocated generally.

4.2.2 Integral Transformations

The nonhomogeneous function $f(x,y)$ in Equations (2.34) and (2.35) may be transformed from its domain integral form to an equivalent set of integrations over the surface when the function $f(x,y)$ is harmonic or biharmonic in the domain V (actually, an n-harmonic function may be treated this way). In this case, either Green's second identity for harmonic functions or the Rayleigh-Green identity for two biharmonic functions is used for the transformation. Consider the case where the function $f(x,y)$ is harmonic in the domain V, $\nabla^2 f(x,y) = 0$. Equation (2.34) can be rewritten using Green's second identity, Equation (2.6), in the following form

$$\int_V (f(x,y)\ \nabla^2 z\ -\ z\ \nabla^2 f(x,y)\ )dV$$

$$= \int_S (f(x,y)\ \frac{\partial z}{\partial n}\ -\ z\ \frac{\partial}{\partial n}\ f(x,y)\ )dS \qquad (4.7)$$

The second term of the domain integration is identically zero. Therefore, if the function $\nabla^2 z$ is set equal to the Green's function G_2, a relationship between the domain integral of Equation (2.34) and a set of equivalent surface integrals is defined. All that is necessary to complete this transformation is the determination of the function z defined by

$$\nabla^2 z(\vec{p},\vec{q}) = \frac{1}{r}\frac{\partial}{\partial r} (r\ \frac{\partial}{\partial r}) = G_2(\vec{p},\vec{q}) \qquad (4.8)$$

Renaming $z(\vec{p},\vec{q})$ to $G_3(\vec{p},\vec{q})$ and solving Equation (4.8) for the new function results in

$$G_3(\vec{p},\vec{q}) = z(\vec{p},\vec{q}) = \frac{|\vec{p}-\vec{q}|^4}{128\pi} (\ln|\vec{p}-\vec{q}| - \frac{3}{2}) \qquad (4.9)$$

The normal derivative of G_3 is calculated as

$$G_3'(\vec{p},\vec{q}) = \frac{|\vec{p}-\vec{q}|^2}{32\pi} (\ln|\vec{p}-\vec{q}| - \frac{5}{4})((x-x_p)n_x + (y-y_p)n_y) \tag{4.10}$$

By substituting Equations (4.9) and (4.10) into Equation (4.7) the domain integrations are transformed into a set of surface integrals of the form

$$\int_V f(x,y)G_2(\vec{p},\vec{q})dV = \int_S (fG_3' - f'G_3)dS \tag{4.11}$$

A similar transformation is found for the domain integral of Equation (2.35) and is defined as

$$\int_V f(x,y)G_1(\vec{p},\vec{q})dV = \int_S (fG_2' - f'G_2)dS \tag{4.12}$$

where the Green's function G_2 is previously defined in Equation (2.10).

Consider the case when the function $f(x,y)$ is biharmonic over the domain V, $\nabla^4 f(x,y) = 0$. In this case, the Rayleigh-Green identity for two biharmonic functions, Equation (2.5), is used to transform Equations (2.34) and (2.35). Therefore, the following form of the identity may be written:

$$\int_V (f(x,y)\nabla^4 w - w\nabla^4 f(x,y))dV = \int_S (f \frac{\partial}{\partial n} (\nabla^2 w)$$

$$- \frac{\partial f}{\partial n} \nabla^2 w + \frac{\partial w}{\partial n} \nabla^2 f - w \frac{\partial}{\partial n} (\nabla^2 f))dS \tag{4.13}$$

The second term of the domain integration of Equation (4.13) is identically zero. If the term $\nabla^4 w$ is set equal to the Green's function G_2, the transformation of the domain integrations of Equation (2.34) into a set of surface integrals is defined. The function $w(p,q)$ is determined for the relationship

$$\nabla^4 w(\vec{p},\vec{q}) = \frac{1}{r} \frac{\partial}{\partial r} (r\frac{\partial}{\partial r} (\frac{1}{r} \frac{\partial}{\partial r}(r \frac{\partial w}{\partial r}))) = G_2(\vec{p},\vec{q}) \tag{4.14}$$

Solving Equation (4.14) for the function $w(p,q)$ and renaming it $G_4(p,q)$ results in

$$G_4(\vec{p},\vec{q}) = \frac{|\vec{p}-\vec{q}|^6}{4608\pi} (\ln|\vec{p}-\vec{q}| - \frac{11}{6}) \tag{4.15}$$

The normal derivative of G_4 is calculated as

$$G_4'(\vec{p},\vec{q}) = \frac{|\vec{p}-\vec{q}|^4}{768\pi} (\ln|\vec{p}-\vec{q}| - \frac{5}{3})((x-x_p)n_x + (y-y_p)n_y) \qquad (4.16)$$

Substituting the above expressions into Equation (4.10) defines the complete transformation for Equation (2.34) into a set of surface integrations as

$$\int_V f(x,y)G_2(\vec{p},\vec{q})dV = \int_S (fG_3' - f'G_3 + G_4'\nabla^2 f - G_4(\nabla^2 f)')dS \qquad (4.17)$$

A similar transformation for the domain integral of Equation (2.35) is

$$\int_V f(x,y)G_1(\vec{p},\vec{q})dV = \int_S (fG_2' - f'G_2 + G_3'\nabla^2 f - G_3(\nabla^2 f)')dS \qquad (4.18)$$

In general, this type of transformation can be continued to any n order harmonic for the function $f(x,y)$. The form of the Green's function and its normal derivative may be written in a general form as (Gipson, Reible, and Savant, 1987)

$$G_k(\vec{p},\vec{q}) = \frac{|\vec{p}-\vec{q}|^{2k-2}}{2^{2k-1} ((k-1)!)^2 \pi} (\sum_{j=1}^{k-1} \frac{1}{j} - \ln|\vec{p}-\vec{q}|) \qquad (4.19)$$

$$G_k'(\vec{p},\vec{q}) = \frac{2(k-1)|\vec{p}-\vec{q}|^{2(k-2)}}{2^{2k-1} ((k-1)!)^2 \pi} (\ln|\vec{p}-\vec{q}| + \frac{1}{2(k-1)}$$

$$- \sum_{j=1}^{k-1} \frac{1}{j}) ((x-x_p)n_x + (y-y_p)n_y) \qquad (4.20)$$

This type of integral transformation eliminates the domain integrations completely for special forms of the function $f(x,y)$. Note that the transformations determined in Equations (4.17) and (4.18) reduce to those defined for the harmonic form of $f(x,y)$. Therefore, this transformation is sufficient to convert both the harmonic and biharmonic forms of the function $f(x,y)$ to a set of surface integrals. Nonetheless, a general function approximated by a finite series can be transformed using the appropriate order of a Green's type identity and the expression in Equations (4.19) - (4.20). Numerical examples for both harmonic and biharmonic types of the function f(x,y) are presented in Chapter 5.

4.2.2.1 Exact Integrations for Special Circumstances with Linear Elements

Exact integration of Equations (2.34) and (2.35) for certain harmonic forms of the function f(x,y) may be determined for a linear boundary element. A numerical formulation is also developed for the general biharmonic form of f(x,y). By predetermining a general form for the harmonic function f(x,y) the transformed domain integrations defined in Equations (2.34) and (2.35) may be performed analytically. The form of the harmonic function is assumed as

$$f(x,y) = C_1xy + C_2x + C_3y + C_4 \tag{4.21}$$

For a linear element analysis, the Cartesian coordinates x and y may be defined using the relationship in Equation (3.3). Substituting the parametrized coordinates into Equation (4.21) redefines f(x,y) as a function of the parameter t, the element constants given in Equation (3.9), and discrete linear end-nodes.

$$f(t) = R_1t^2 + R_2t + R_3 \tag{4.22}$$

where R_1, R_2, and R_3 are defined as

$$R_1 = C_1A_xA_y \qquad R_2 = C_1(A_xy_i + A_yx_i) + C_2A_x + C_3A_y$$

$$R_3 = C_1x_iy_i + C_2x_i + C_3y_i + C_4 \tag{4.23}$$

In a similar manner, the normal derivative of the general harmonic function, defined in Equation (4.21), is calculated as

$$\frac{\partial f(t)}{\partial n} = S_1t + S_2 \tag{4.24}$$

where S_1 and S_2 are defined as

$$S_1 = C_1(A_yn_x + A_xn_y) \qquad S_2 = (C_1y_i + C_2)n_x + (C_1x_i + C_3)n_y \tag{4.25}$$

Substituting this special form of f(x,y) into Equation (4.8) results in the following form of the required surface integrations

$$\int_v f(x,y)G_2(\vec{p},\vec{q}) \, dV = \frac{1_e}{64\pi} \sum_{}^{e} \int_0^1 (E(R_1 t^2 + R_2 t$$

$$+ R_3)X(\ln X - \frac{5}{2}) - \frac{1}{4}(S_1 t + S_2)X^2(\ln X - 3))dt \qquad (4.26)$$

where the position vector magnitude X and the element constant E are given in
Equations (3.8) and (3.14), respectively. The integration table defined in
Equations (3.15) - (3.24) and complemented by Equations (3.83), (3.74), (3.123)
and (3.124) provides all the necessary components to define the analytic expression for the integration as a summation over n linear elements:

$$\int_v f(x,y)G_2(\vec{p},\vec{q})dV = \frac{1_e}{64\pi} \sum_{}^{e} (-\frac{S_1 P_1}{4} L_5$$

$$+ (R_1 AE - \frac{1}{4}(S_1 P_2 + S_2 P_1))L_4$$

$$+ (E(R_1 B + R_2 A) - \frac{1}{4}(S_1 P_3 + S_2 P_2))L_3$$

$$+ (E(R_1 C + R_2 B + R_3 A) - \frac{1}{4}(S_1 P_4 + S_2 P_3))L_2$$

$$+ (E(R_2 C + R_3 B) - \frac{1}{4}(S_1 P_5 + S_2 P_4))L_1$$

$$+ (ER_3 C - \frac{1}{4} S_2 P_5)L_0 - \frac{5E}{2}(R_1 T_3 + R_2 T_2 + R_3 T_1)$$

$$+ \frac{3}{4}(S_1 U_2 + S_2 U_1)) \qquad (4.27)$$

where

$$P_1 = A^2 \qquad P_2 = 2AB \qquad P_3 = 2AC + B^2$$

$$P_4 = 2BC \qquad P_5 = C^2 \qquad T_n = \frac{A}{n+2} + \frac{B}{n+1} + \frac{C}{n} \qquad (4.28)$$

$$U_n = \frac{P_1}{n+4} + \frac{P_2}{n+3} + \frac{P_3}{n+2} + \frac{P_4}{n+1} + \frac{P_5}{n}$$

The element constants A, B, and C are previously defined for a linear element in Equation (3.9). The analytical expression for the domain integral of Equation (2.35) based on the same set of element constants, the complete integration table, and the constants of Equation (4.28), is calculated as

$$\int_v f(x,y)G_1(p,q)\ dV = \frac{l_e}{8\pi} \sum^e \left(-\frac{S_1 A}{2} L_3 \right.$$

$$+ (R_1 E - \tfrac{1}{2}(S_1 B + S_2 A))L_2$$

$$+ (R_2 E - \tfrac{1}{2}(S_1 C + S_2 B))L_1 + (R_3 E - \tfrac{1}{2} S_2 C)L_0$$

$$\left. - E(\frac{R_1}{3} + \frac{R_2}{2} + R_3) + S_1 T_2 + S_2 T_1 \right) \tag{4.29}$$

The analytical expressions for the domain integrals of Equations (2.34) and (2.35) provide a very accurate technique to work a wide range of nonhomogeneous biharmonic problems. Although the exact analysis is restricted to functions of the form given in Equation (4.21), any general biharmonic function may be transformed and evaluated numerically.

4.2.2.2 Integration Strategy for a General Parametric Element

Equations (4.17) and (4.18) may be rewritten to accommodate any type or combination of elements. In Chapter 3, an extensive analysis of linear and high order elements was presented. By substituting the appropriate form of the position vector and the Jacobian for the desired element into Equations (4.17) and (4.18), the domain integral is expressed in terms of a series of parametric surface integrations. The general forms of the integrations for any element type are given as

$$\int_v f(x,y)G_2(\vec{p},\vec{q})\ dV = \sum^e \int_0^1 (fG_3' - f'G_3$$

$$+ \nabla^2 f G_4' - G_4(\nabla^2 f)')\ |J|\ dt \tag{4.30}$$

$$\int_V f(x,y) G_1(\vec{p},\vec{q}) \ dV = \sum^e \int_0^1 (\ fG_2' - f'G_2$$

$$+ \ \nabla^2 fG_3' - G_3(\nabla^2 f)' \ ) \ |J| \ dt \qquad\qquad (4.31)$$

where the summation is over the number of elements, e, used to describe the discrete surface.

4.2.3 Domain Fanning

The nonhomogeneous terms of the integral equations defined in Equations (2.28) and (2.29) must be evaluated over the region V, the problem domain. The evaluation of these domain integrations may be handled in a number of ways. A popular technique is the use of internal cells where the domain is subdivided into a series of volume elements over each of which a numerical quadrature scheme is applied. This type of procedure requires a discretization of the problem domain and can be difficult to implement for a general region. Monte Carlo quadrature techniques, which have been successfully used to evaluate domain integrations associated with the Poisson equation (Gipson, 1985; Gipson and Camp, 1985), do not require domain discretization. The fundamental disadvantage of Monte Carlo integration is that its accuracy increases only on the order of the inverse square root of the number of sampling points. Neither the volume cell nor the Monte Carlo method is intrinsically sensitive to the singular nature of the fundamental solution associated with the domain terms of Equations (2.28) and (2.29) near the source point.

In this work, an improved domain quadrature technique similar to that used by Telles (1983) is implemented. The method combines the convenience of higher order numerical integration over a triangular area with the inherent advantages of nondiscretization of the domain. This technique divides the domain into a series of triangular areas, each formed implicitly by a set of three vertices; two are consecutive discrete nodal points of a boundary element describing the

surface of the domain and the other is the source point (x_p, y_p) under consideration. Each of the elemental triangular regions is divided into a series of smaller triangular areas, as shown in Figure 4.1. The effect is to concentrate quadrature points in a region close to the source point (x_p, y_p) where the Green's
function is singular, and relax the intensity of the quadrature in areas where
the function is more well-behaved. Each of the domain integrals defined in Equations (2.34) and (2.35) is referenced to a discrete source point. By maintaining
the source point as a vertex of the triangular area, the integration scheme
automatically distributes its quadrature points in a way that is sensitive to the
singularity of the Green's functions as the other two vertices move from element
to element around the boundary. The effect is to "fan" the region about the
point in question. Another advantage of this method is that it is equally ap-
plicable to source points on the interior of the domain. This feature is important in the evaluation of the domain integrals of Equations (2.11) and (2.12)
which define the internal solution of the field variables.

4.3 Iterative Solution

If the nonhomogeneous source term $f(x,y)$ is also a function of the field
variables and their derivatives, the solution to the governing coupled boundary
integral expressions, Equations (2.28) and (2.29), is obtained by an iterative
technique. The first iteration solves the homogeneous form of the governing
equations and uses that solution to update the domain source terms, Equations
(2.34) and (2.35), for the second iteration. After all the source terms are
calculated, Equations (2.28) and (2.29) are solved to determine an intermediate
solution of ψ, ψ', ω, and ω'. The updated solution for each preceding iteration
is obtained by relaxing the intermediate solution and adding that to either the
homogeneous solution or to the previous intermediate solution. For example, the
$k+1$ iteration for ψ and ω would be

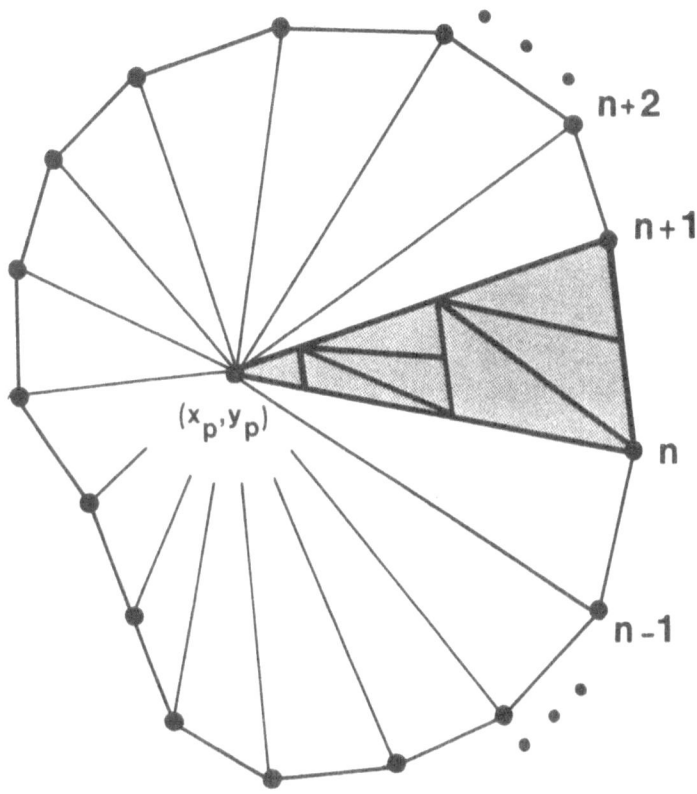

Figure 4.1 Distribution of Triangular Quadrature Regions for the "Fanning"
            Domain Integration Technique Related to a Particular Source
            Point. The Shaded Area Isolates the Distribution Over a
            Single Elemental Triangle.

$$\psi^{k+1} = \theta \, \psi^k + (1 - \theta) \, \psi^{k+1}$$

$$\omega^{k+1} = \theta_1 \omega^k + (1 - \theta_1) \, \omega^{k+1}$$

(4.32)

where ψ^k and ω^k are the solutions to the kth iteration or the homogeneous form of Equations (2.28) and (2.29) and θ and θ_1 are the appropriate relaxation factors. The iteration procedure is continued until a suitable convergence criterion is satisfied.

Evaluation of the terms in Equations (2.34) and (2.35) for an iterative solution will be accomplished using the domain "fanning" integration technique discussed earlier. The form of the function f may involve the field variables ψ and ω and their derivatives in any combination. The "fanning" integration scheme requires the values of these functions at every quadrature point in the domain. In this work, the form of the function f will be restricted to functions of the form ψ, ω, $d\psi/dx$, $d\psi/dy$, $d\omega/dx$, and $d\omega/dy$. Values of these functions are automatically calculated at an array of uniformly distributed points in the domain and are combined with the boundary solution to create a series of solution maps, one for each of the six functions. When the integration scheme calls for a value of a particular function at an arbitrary triangular quadrature point, the point in question is located geometrically within each solution map array. Between two and four points in closest proximity to the quadrature point are located. The value of the function is then determined through linear interpolation of the set of field values associated with these points in the array map.

The locations of the points in each solution map are implicitly defined within the formulation. Cartesian coordinates of each array point are defined in terms of the maximum and minimum spatial coordinates of the problem domain under consideration. Dividing the difference between the maximum and minimum points in each coordinate direction by a prescribed number of division defines the solution map point spacing, as shown in Figure 4.2. Each point is checked to determine if it actually lies within the domain using a residue theorem technique (Gipson,

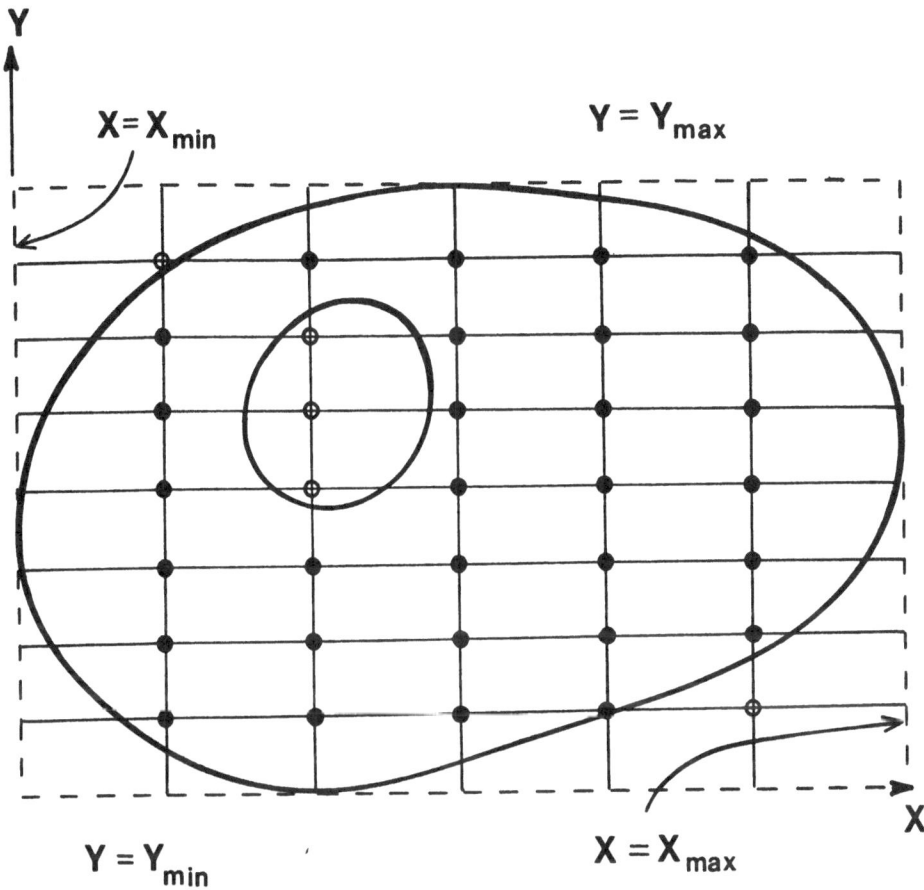

● Points inside the domain

⊕ Points outside the domain

Figure 4.2   Internal Point Locations for a Solution Map for an Arbitrary Domain

1986). When an array point is found to be outside the domain, it is not considered in either the "fanning" integration or the solution map interpolation techniques.

The advantage of the solution map scheme is that it requires relatively few
internal point calculations for the values of the functions ψ, ω, and their
derivatives. The execution time of the technique is much faster than it would be
if the function were calculated at every triangular quadrature point. Any additional error introduced by the interpolation algorithm is minimal and offset by
the decreases in run time. Overall, an iterative solution based on the solution
map technique performs consistently and accurately. Several examples of iterative solutions for various combinations of the functions ψ, ω and their derivatives are presented in Chapter 5.

4.4   Internal Point Calculations

Boundary element formulations generally consist of a series of integrations
over the surface of the problem domain. The values of the field variables at any
points interior to the surface are calculated from a set of surface integrations
that require a "complete" solution on the boundary. For a biharmonic analysis,
the integral relationships that defined the values of the functions ψ and ω at
any internal point are given in Equations (2.11) and (2.12). The first derivatives of ψ and ω with respect to both Cartesian coordinates x and y were
previously defined in Equations (2.15), (2.16), (2.21), and (2.22). If linear
isoparametric elements are used to describe the surface geometry, the integrations necessary to calculate the values of ψ, ω, and their derivatives may be
performed analytically. Subparametric forms of both the quadratic and the Overhauser elements may also be evaluated analytically if the geometry is piecewise
linear. Integrations over curved geometries using quadratic or Overhauser
formulations may be performed using any appropriate numerical quadrature.

4.4.1 Exact Integration Strategy for Linear Elements

Evaluation of the functions ψ and ω at any internal point in the domain V are calculated using Equations (2.28) and (2.29), respectively. The integrations defined in these equations are determined through analytical expressions derived in Chapter 3. Since the source point (x_p, y_p) for an internal point calculation is never located on the element itself, the logarithmic singularity encountered previously is no longer a concern. Therefore, the appropriate components of the exact expression for the required integrations are given in Equations (3.25) - (3.32).

The values of the first derivatives of the functions ψ and ω are calculated from the relationships in Equations (2.15), (2.16), (2.21), and (2.22). The derivative operator acts exclusively upon the Green's functions as shown in Equations (2.17) - (2.20) and (2.23) - (2.26). The field variables may be approximated by a series of discrete nodal values and a corresponding shape function set. Substitution of the linear shape functions previously defined in Chapter 3 into the expressions defining the derivatives results in integrands of the form t^n, t^n/X, t^n/X^2, and $t^n \ln X$. Only the integrations involving terms of the form t^n/X^2 have not been previously defined. Therefore, the following additions to the integration table are given as

$$M_0 = \int_0^1 \frac{dt}{X^2} = \left(\frac{2A+B}{A+B+C} - \frac{B}{C} + 2AI_0 \right)/\Delta \qquad \Delta > 0$$

$$= \frac{1}{3A^2} \left(\left(\frac{2A}{B}\right)^3 - \frac{1}{\left(1 + \frac{B}{2A} \right)^3} \right) \qquad \Delta = 0$$

$$(4.33)$$

$$M_1 = \int_0^1 \frac{t}{X^2} \, dt = -\left(\frac{2C+B}{A+B+C} - 2BI_0 \right)/\Delta \qquad \Delta > 0$$

$$= -\left(\frac{1}{2} + \frac{B}{12A} \right)/\left(1 + \frac{B}{2A} \right)^3 + \frac{2A^2}{3B^2} \qquad \Delta = 0$$

$$(4.34)$$

where $\Lambda = 4AC - B^2$. Also,

$$M_2 = \int_0^1 \frac{t^2}{x^2} \, dt = - \frac{1}{A(A+B+C)} + \frac{C}{A} M_0 \tag{4.35}$$

The parametrized linear element formulation presented in Chapter 3 is corrected for analytical derivative calculations by substituting the following terms into Equations (3.25) - (3.28).

$$H_j = 2(m_n I_1 - 2E(A_n M_2 + B_n M_1))$$

$$H_i = 2(m_n I_0 - 2E(A_n M_1 + B_n M_0)) - H_j \tag{4.36}$$

$$G_j = A_n I_2 + B_n I_1$$

$$G_i = A_n I_1 + B_n I_0 - G_j \tag{4.37}$$

$$L_j = 2(2E(A_n I_2 + B_n I_1) + m_n(L_1 - \frac{1}{2}))$$

$$L_i = 2(2E(A_n I_1 + B_n I_0) + m_n(L_0 - 1)) - L_j \tag{4.38}$$

$$K_j = A_n L_2 + B_n L_1 - (\frac{A_n}{3} + \frac{B_n}{2})$$

$$K_i = A_n L_1 + B_n L_0 - (\frac{A_n}{2} + B_n) - K_j \tag{4.39}$$

where n = 1, 2 corresponds to the x derivative or the y derivative formulation, respectively. The constants A_n, B_n, and m_n are given as

$$A_1 = A_x \qquad\qquad A_2 = A_y$$

$$B_1 = B_x \qquad\qquad B_2 = B_y \tag{4.40}$$

$$m_1 = n_x \qquad\qquad m_2 = n_y$$

Domain integrations required in the nonhomogeneous form of the equations used to calculate derivatives at internal points are evaluated analytically for a special form of the function $f(x,y)$, given in Equation (4.21). By substituting the appropriate form of the Green's functions and the special version of $f(x,y)$ into Equations (4.11) and (4.12) domain integrals are transformed into a series of boundary integrations. If the surface is decribed by linear elements, the transformed analytic relationships over each element are defined as

$$
\int_0^1 f(x,y)\, \frac{\partial G_2}{\partial x_i}\, dV = \frac{1}{32\pi}\, \sum^e \left(-\frac{Q_1 L_4}{2} + \left(EP_1 - \frac{Q_2}{2}\right) L_3 \right.
$$

$$
+ \left(EP_2 - \frac{Q_3}{2}\right) L_2 + \left(EP_3 - \frac{Q_4}{2}\right) L_1 + \left(EP_4 - \frac{Q_5}{2}\right) L_0
$$

$$
- \frac{3E}{2} \left(\frac{P_1}{4} + \frac{P_2}{3} + \frac{P_3}{2} + P_4 \right)
$$

$$
\left. + \frac{5}{4} \left(\frac{Q_1}{5} + \frac{Q_2}{4} + \frac{Q_3}{3} + \frac{Q_4}{2} + Q_5 \right) \right)
$$

(4.41)

$$
\int_v f(x,y)\, \frac{\partial G_1}{\partial x_i}\, dV = \frac{1}{8\pi} \sum^e \left(2E(P_1 I_3 + P_2 I_2 + P_3 I_1 + P_4 I_0) \right.
$$

$$
+ (R_1 n_i - S_1 A_i) L_2 + (R_2 n_i - (S_1 B_i + S_2 A_i)) L_1
$$

$$
+ (R_3 n_i - S_2 B_i) L_0 - n_i \left(\frac{R_1}{3} + \frac{R_2}{2} + R_3 \right) + \frac{1}{3} S_1 A_i
$$

$$
\left. + \frac{1}{2} \left(S_1 B_i + S_2 A_i \right) + S_2 B_i \right)
$$

(4.42)

where the function constants R_1, R_2, R_3, S_1, and S_2 are previously defined in Equations (4.23) - (4.25). The remaining constants are given as

$$
P_1 = R_1 A_i \qquad\qquad P_2 = R_1 B_i + R_2 A_i
$$

$$
P_3 = R_2 B_i + R_3 A_i \qquad\qquad P_4 = R_3 B_i
$$

(4.43)

$$Q_1 = AS_1 A_i - AR_1 n_i$$

$$Q_2 = S_1(AB_i + BA_i) + AS_2 A_i - n_i(AR_2 + BR_1)$$

$$Q_3 = S_1(BB_i + CA_i) + S_2(AB_i + BA_i)$$

$$- n_i(AR_3 + BR_2 + CR_1)$$

(4.44)

$$Q_4 = CS_1 B_i + S_2(BB_i + CA_i) - n_i(BR_3 + CR_2)$$

$$Q_5 = CS_2 B_i - n_i CR_3$$

where the subscript "i" is equal to 1 or 2 corresponding to the x or y derivative, respectively. Therefore, the constants A_i, B_i, and the direction cosine n_i are given as

$$A_1 = A_x \qquad\qquad A_2 = A_y$$

$$B_1 = B_x \qquad\qquad B_2 = B_y \qquad\qquad (4.45)$$

$$n_1 = n_x \qquad\qquad n_2 = n_y$$

Although the analytical expressions defined in Equations (4.41) and (4.42) may seem cumbersome, they are easily calculated, require less time to execute, and are more accurate than integration using numerical quadrature. The analytical analysis of the simple form of the function $f(x,y)$ is justified since it has many applications in a wide range of engineering problems.

The domain integrals for nonhomogeneous analysis may be evaluated for the special case when $\nabla^4 f(x,y) = 0$ by the transformation technique discussed earlier. Transformation of the domain integrals into surface integrations requires spatial derivatives for the functions G_2, G_2', G_3, G_3', G_4, G_4'. The x and y derivatives of G_2 and G_2' are defined in Equations (2.19) - (2.20) and (2.25) - (2.26). The remaining x-coordinate derivatives are calculated as

$$\frac{\partial G_3}{\partial x_p} = \frac{1}{64\pi}\left((x-x_p) \ X \ \left(\ln X - \frac{5}{2} \right) \right) \qquad (4.46)$$

$$\frac{\partial G'_3}{\partial x_p} = \frac{1}{64\pi} \left(2(x-x_p) \ Z \ \left(\ln X - \frac{3}{2} \right) \right.$$

$$\left. + n_x X \left(\ln X - \frac{5}{2} \right) \right) \tag{4.47}$$

$$\frac{\partial G_4}{\partial x_p} = \frac{1}{1536\pi} \left((x-x_p) \ x^2 \left(\ln X - \frac{10}{3} \right) \right) \tag{4.48}$$

$$\frac{\partial G'_4}{\partial x_p} = \frac{1}{1536\pi} \left(4(x-x_p) \ Z \ X \ \left(\ln X - \frac{17}{6} \right) \right.$$

$$\left. + n_x x^2 \left(\ln X - \frac{10}{3} \right) \right) \tag{4.49}$$

The necessary y-coordinate derivatives are determined as

$$\frac{\partial G_3}{\partial y_p} = \frac{1}{64\pi} \left((y-y_p) \ X \ \left(\ln X - \frac{5}{2} \right) \right) \tag{4.50}$$

$$\frac{\partial G'_3}{\partial y_p} = \frac{1}{64\pi} \left(2(y-y_p) \ Z \ \left(\ln X - \frac{3}{2} \right) \right.$$

$$\left. + n_y X \left(\ln X - \frac{5}{2} \right) \right) \tag{4.51}$$

$$\frac{\partial G_4}{\partial y_p} = \frac{1}{1536\pi} \left((y-y_p) \ x^2 \left(\ln X - \frac{10}{3} \right) \right) \tag{4.52}$$

$$\frac{\partial G'_4}{\partial y_p} = \frac{1}{1536\pi} \left(4(y-y_p) \ Z \ X \ \left(\ln X - \frac{17}{6} \right) \right.$$

$$\left. + n_y x^2 \left(\ln X - \frac{10}{3} \right) \right) \tag{4.53}$$

where the variable Z is

$$Z = (x-x_p)n_x + (y-y_p)n_y \tag{4.54}$$

For a general form of the function $f(x,y)$ the "fanning" domain integration technique is used to approximate the integrals required for the calculation of the derivatives of ψ and ω at internal points. Although the execution time is dramatically increased, the resulting solution is very accurate and generally consistent for well behaved functions. However, if the rate of change of the function f(x,y) becomes large over a small area, the order of the quadrature over that region should be appropriately increased. Refining the boundary discretization may improve the integration, since the number of quadrature points directly corresponds to the number of elements.

4.4.2 Integration Strategy for Higher Order Elements

The correct form of the integrations required to calculate the values of the field variables ψ, ω, and their derivatives at any internal point for a general element are easily obtained. By substituting the appropriate forms of the shape functions, the Jacobian transformation, and the position vector into Equations (2.11), (2.12), (2.15), (2.16), (2.21), and (2.22), the required forms integral expressions are determined. The reader is referred to Chapter 3 for the development of these element parameters for both the quadratic and the Overhauser elements.

In integrating over the boundary for a general internal point, the position vector is always non- zero. Therefore, no form of special quadrature is necessary since the logarithmic Green's function is no longer singular. A one-dimensional Gaussian quadrature over the parameterized element is used in the calculation of the values of the field variables and their derivatives at any internal point.

If the geometry is piecewise linear or assumed linear, the integrals required in internal point calculations may be evaluated analytically. The procedure is identical to that presented in Chapter 3. The Jacobian becomes a constant and is factored from the integrations. The resulting integrals have components defined in the integration table developed in Chapter 3 and supplemented in this chapter. For a quadratic element, the required integrations for the calculation of functions ψ and ω at any internal point are defined in Equations (3.85) - (3.92). The corresponding set of relationships for an Overhauser element are given in Equations (3.125) - (3.132). Surface integrations defining the values of the derivatives of ψ and ω are also calculated analytically for the subparametric versions of both the quadratic and the Overhauser elements. However, the necessary domain integrations involving the function f(x,y) are evaluated numerically using either the integration transformation technique or the domain "fanning" scheme.

4.5 Summary and Concluding Remarks

Two methods for handling domain integrations have been presented. Integrations over the domain may be tranformed for harmonic and biharmonic forms of the function f(x,y) into a set of surface integrals. Exact analysis for special cases of the resulting surface integrations were derived. The superior accuracy of an exact formulation may offset any error induced by assuming a linear variation of the geometry. The domain "fanning" technique provided implicit volume quadrature for forms of the function f(x,y) which cannot be transformed. Although evaluated numerically, the domain "fanning" method has inherent sensitivity to the distribution of the Green's function resulting in accurate and consistent solutions.

Internal point calculations for the values of the field variables over several types of elements were defined. An exact expression of an isoparametric linear element and the subparametric versions of the quadratic and the Overhauser element were derived. The derivatives of the field variables were evaluated numerically for all element types. However, derivative calculations for a special form of the function f(x,y) over a linear element were performed analytically. Numerical calculations were very accurate as long as the point in question remained outside a zone measured by approximately half the element length from the boundary. Inside this zone the calculations became inaccurate. In Chapter 5, several numerical examples will be worked demonstrating some of the various methods presented in this chapter.

CHAPTER 5

EXAMPLE ANALYSES

5.1 Introduction

In Chapter 6, a computer program will be presented for the analysis of the
nonhomogeneous biharmonic equation. In this chapter are presented actual case
studies using this program and a more elaborate version of the same program for
working the nonlinear and iterative problems. The example problems presented
here demonstrate the versatility of the formulation developed in Chapters 3 and 4
over a wide range of engineering problems and also illustrate practical modeling
techniques for boundary element analysis.

The example problems studied may be divided into two categories. The first
will be a myriad of problems involving various forms of governing equations of
thin plates with small deflections. The second category will be a study of incompressible viscous fluid flow at low Reynolds numbers. In each case, the
governing physical process will be identified along with the engineering applications. All example analyses are compared to existing analytical solutions or
current published numerical approximations when available. The availability of
analytical solutions for small deflections of thin plates of simple geometries
provides an excellent base upon which to compare the different element types
developed in the preceding chapters. Once an element hierarchy is established,
most of the proceeding examples will use an element type determined to produce
superior results.

5.2 Deflections of Thin Plates

The first category of problems to be studied has to do with the small deflections of thin plates. A thin plate is an initially flat structural element where the ratio of the thickness, measured normal to the midplane, to the smallest span dimension is less than 1/20 (Ugural, 1981). Unless otherwise denoted, the examples presented will involve thin plates composed of homogeneous isotropic materials. A homogeneous "plate" body has identical elastic properties throughout the material. If the material properties are also equivalent in all directions, the material is isotropic. The governing equation for the deflection of a thin structural plate under transverse loading $P(x,y)$, first derived by Lagrange in 1811, is given as

$$\nabla^4 w = \frac{P(x,y)}{D} \tag{5.1}$$

where w is the midplane deflection and D is the flexural rigidity. The resulting nonhomogeneous biharmonic equation requires that two boundary conditions of the form given in Equation (2.2) be satisfied on each edge.

A clamped or built-in edge condition requires that both the deflection, w, and the slope, $\partial w/\partial n$, be equal to zero at the boundary. This type of edge condition matches directly with the "forced" boundary conditions of the integral representation of the governing equation. Therefore, a clamped or built-in edge may be modeled for any type of geometry.

The second type of boundary conditions allowable with this analysis is a simply supported edge. In this case, the deflection and the normal bending moment, M_n, are both zero. The deflection condition is directly compatible to the "forced" conditions of the governing equation. However, the bending moment edge condition must be converted into a form which matches one of the remaining three boundary conditions of Equation (2.2). The normal bending moment, M_n, is defined as

$$M_n = - D \left(\frac{\partial^2 w}{\partial n^2} + \nu \frac{\partial^2 w}{\partial s^2} \right) \tag{5.2}$$

where ν is Poisson's ratio and the coordinate s is measured tangent to the surface at any point. For any polygonal shape, along each rectilinear simply supported edge of the boundary, the term $\partial^2 w/\partial s^2$ is identically zero. Observing that M_n is specified as zero on a simply supported edge, the remaining term of Equation (5.2), $\partial^2 w/\partial n^2$, also vanishes. The moment function M is defined as

$$M = \frac{M_x + M_y}{1+\nu} = \frac{M_n + M_s}{1+\nu} \tag{5.3}$$

where M_x and M_y are the bending moments in Cartesian coordinates and M_s is the tangential bending moment. Rewriting the moment function in terms of the deflection results in the following:

$$M = - D \left(\frac{\partial^2 w}{\partial x^2} + \frac{\partial^2 w}{\partial y^2} \right) = - D \left(\frac{\partial^2 w}{\partial n^2} + \frac{\partial^2 w}{\partial s^2} \right) \tag{5.4}$$

Therefore, on any simply supported rectilinear edge of a polygonal shape the moment function is identically zero (Timoshenko and Woinowsky-Krieger, 1959). This relationship may be recast in the form of a "forced" boundary condition as

$$\nabla^2 w = - \frac{M}{D} = 0 \tag{5.5}$$

For the integral formulation developed in this work, a simply supported edge condition is possible for any polygonal shape and is specified by prescribing both the deflection, w, and the Laplacian of the deflection, $\nabla^2 w$, as zero along the boundary.

Numerical solutions will be presented for several different types of transverse loadings on thin plates of various geometries under two types of support conditions. Circular plate analyses will be limited to those with clamped supports, whereas both clamped and simply supported end conditions will be analyzed possible for polygonal shapes. Numerical quadrature will be used in evaluating the necessary boundary integrals for all elements, except linear elements, for any curved geometry. All polygonal shapes will be analyzed with the analytical expressions developed for the subparametric version of each element. In each case, the various element types will be compared and their performance evaluated.

5.2.1 Circular Clamped Plates

Various axisymmetric loadings which only depend upon the radial coordinate will be considered. The governing equation for the deflection in terms of the radial coordinate is given as

$$\nabla^4 w = (\ \frac{d^2}{dr^2} + \frac{1}{r}\frac{d}{dr}\ )\ (\ \frac{d^2 w}{dr^2} + \frac{1}{r}\frac{dw}{dr}\ ) = \frac{P(r)}{D} \tag{5.6}$$

where P(r) is the transverse loading function. The outside radius, a, of all the examples will be set equal to the numerical value of two. The clamped boundary conditions have been established and are applied as previously discussed.

5.2.1.1 Concentrated Load at Plate Center.

The deflection and moment function for a clamped circular plate with a concentrated load at its center are calculated. A concentrated load is one of the simplest loading conditions for any type of boundary element formulation. Since a concentrated load acts at a point, the loading function P(r) may be replaced by the value of the concentrated load multiplied by the Dirac delta function. Substituting this form of the loading into the necessary domain integrals results in a single evaluation of the integrand at the location of the concentrated load. Therefore, any error in the solution may be attributed directly to the surface integrations of the governing integral equation. This problem will provide an excellent format for the comparison and the evaluation of the three isoparametric elements presented in Chapter 3.

A series of boundary element meshes were used to discretize the circular problem domain. Each of the three elements; linear, quadratic, and Overhauser, are used to describe each mesh. An illustration of the ability of each element

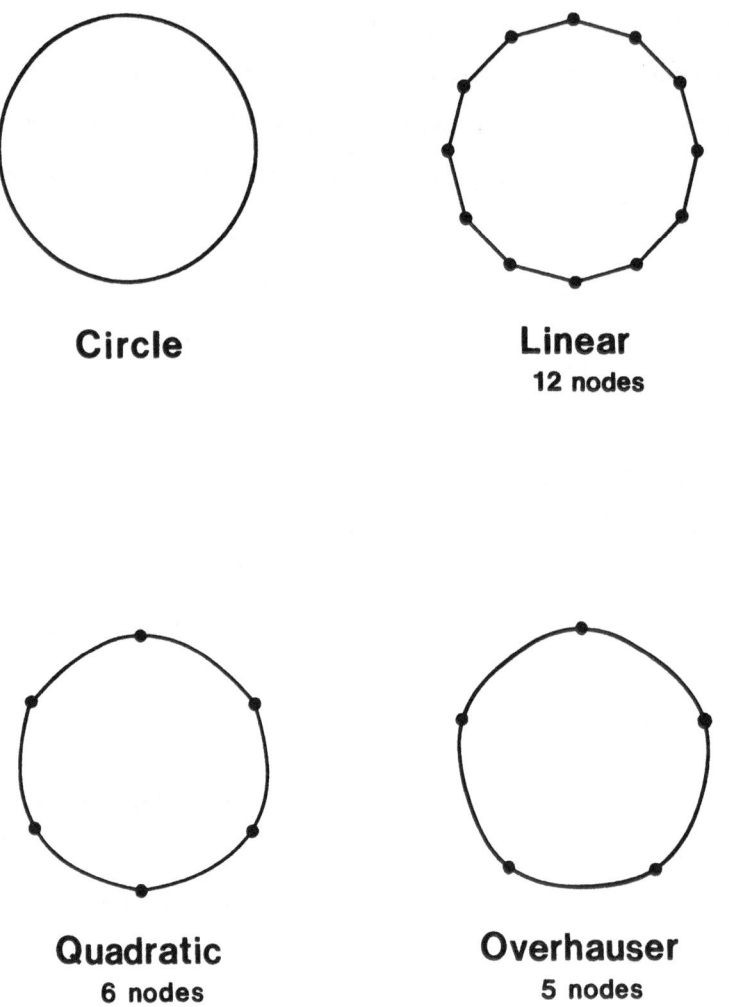

Circle

Linear
12 nodes

Quadratic
6 nodes

Overhauser
5 nodes

Figure 5.0   Comparison of Example Models Using Various
Element Types to a Circle.

type to accurately represent the circular geometry is shown in Figure 5.0. The Overhauser elements are used to model the surface as a C_1 continuous curve, whereas the linear and quadratic element representations of the boundary have discontinous derivatives between elements. To allow for a fair comparison, the number of discrete nodal points and their locations were held constant for each mesh. The absolute percentage error between the boundary element solution and the exact solution, given by Timoshenko and Woinowsky-Krieger (1959), for the center deflection is shown in Figure 5.1. Several interesting observations may be made from this graph. Even though the linear element analysis was performed using analytical expressions for the integrations, the percentage error is much greater than that in both the quadratic and Overhauser element analyses. A solution using just six nodes and six Overhauser elements deviated only 2% from the exact value for the center deflection. Both the quadratic and Overhauser analyses provided excellent solutions when ten or more elements were used. The result of this comparison seems to indicate that the Overhauser element is superior to both the linear and quadratic elements for curved geometries. A boundary mesh using 20 Overhauser elements to describe the circle was used to calculate the deflection, w, and the moment function, M, at several internal points. The results are listed in Table 5.1.

5.2.1.2 Uniform Load. Consider an extension of the previous example to that of a plate carrying a uniform transverse load. The domain integrations are transformed into surface integrals using the techniques discussed in Chapter 4. Since the function $P(r) = q$ is a constant, the transformation converts each of the necessary domain integrations into a single corresponding boundary integral. A comparison of the three element types for this problem is shown in Figure 5.2. The solution obtained from the analytical formulation using linear elements is extremely poor when using less than about 30 nodes. However, the quadratic and the Overhauser elements provide outstanding solutions using only ten elements.

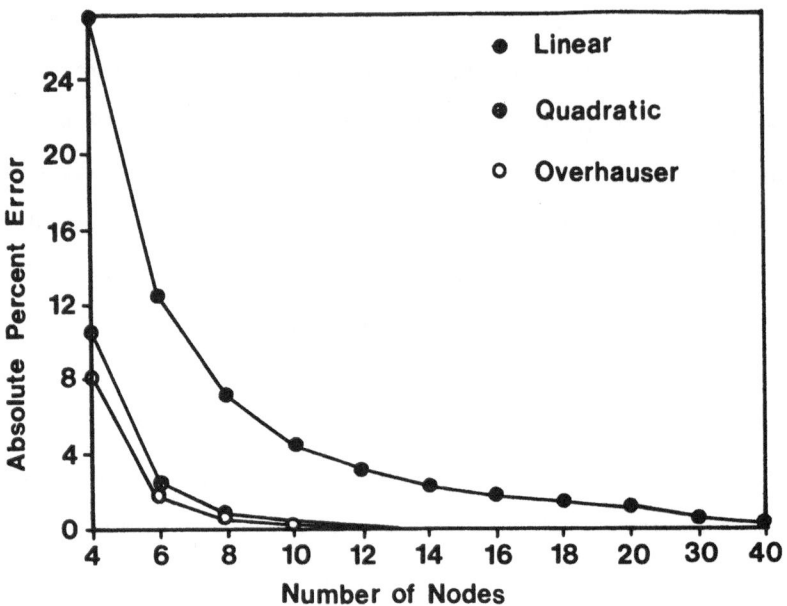

Figure 5.1   Absolute Percent Error For the Center Deflection of a Clamped Circular Plate With a
Concentrated Load at the Center.

TABLE 5.1

Deflection and Moment Function for a Clamped Circular
Plate with a Concentrated Load P at its Center.

r/a	α	α exact	β	β exact
0.0	0.01989	0.01989	-	-
0.2	0.01653	0.01654	0.17655	0.17657
0.4	0.01087	0.01088	0.06624	0.06625
0.6	0.00541	0.00542	0.00171	0.00172
0.8	0.00147	0.00148	-0.04407	-0.04406

Note: Deflection, $w = \alpha P a^2/d$, Moment Function $M = \beta P$, and
radius a.

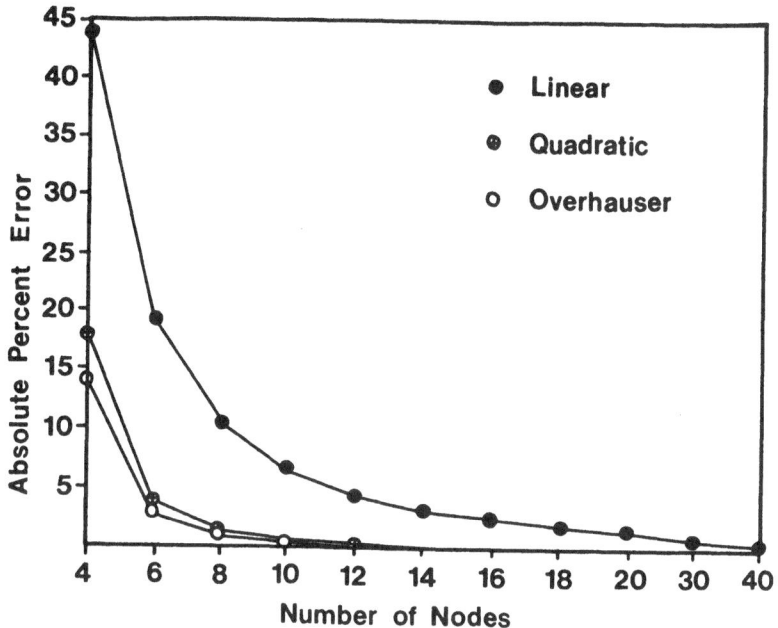

Figure 5.2 Absolute Percent Error For the Center Deflec-
            tion of a Clamped Circular Plate Under a
            Uniform Load.

TABLE 5.2

Deflection and Moment Function for a Clamped Circular
           Plate Under a Uniform Load P(r)=q.

r/a	α	α exact	β	β exact
0.0	0.01562	0.01562	0.12499	0.12500
0.2	0.01439	0.01440	0.11499	0.11500
0.4	0.01102	0.01102	0.08499	0.08500
0.6	0.00640	0.00640	0.03499	0.03500
0.8	0.00202	0.00202	-0.03501	-0.03500

Note: Deflection, $w=\alpha qa^4/D$, Moment Function
       $M=\beta qa^2$, and radius a.

Table 5.2 lists the deflection, w, and moment function, M, at several internal points determined from an analysis using 20 Overhauser elements.

5.2.1.3 <u>Quadratic Load</u>. A transverse loading of the form $P(r) = q(r/a)^2$ over a clamped circular plate is presented in this example. The domain integrations representing the load are converted into a series of surface integrals by the biharmonic version of the integral transformations discussed in Chapter 4. For this particular loading, each domain integration is transformed into an equivalent set of three surface integrals. The three element types are once again compared for several different boundary discretizations. The results of this analysis are shown in Figure 5.3 in terms of the absolute percentage error between the numerical solution and the exact value for the center deflection. As the loading function $P(r)$ increases in order, the number of nodes required for an acceptable solution also increases. In Table 5.3, a solution using 20 Overhauser elements is presented for the deflection and moment function at several internal points.

The three preceding examples demonstrated several important features of the boundary integral formulation presented in this work. Many forms of the loading function $P(r)$ may be rewritten as surface integrals avoiding any type of domain quadrature. The shape functions associated with quadratic and Overhauser elements not only represent the geometry better than a linear element, but also provide a much more accurate solution with far fewer nodes. The decrease in execution time attributed to the analytic expression used in the linear element formulation does not compensate for its lower order approximation of the field variables. However, if a very large number of nodes are required for a particular problem, the difference in the solution obtained using any of the three element types is negligible. In this case the linear element formulation displays a slight advantage over the other two higher order elements in total execution time.

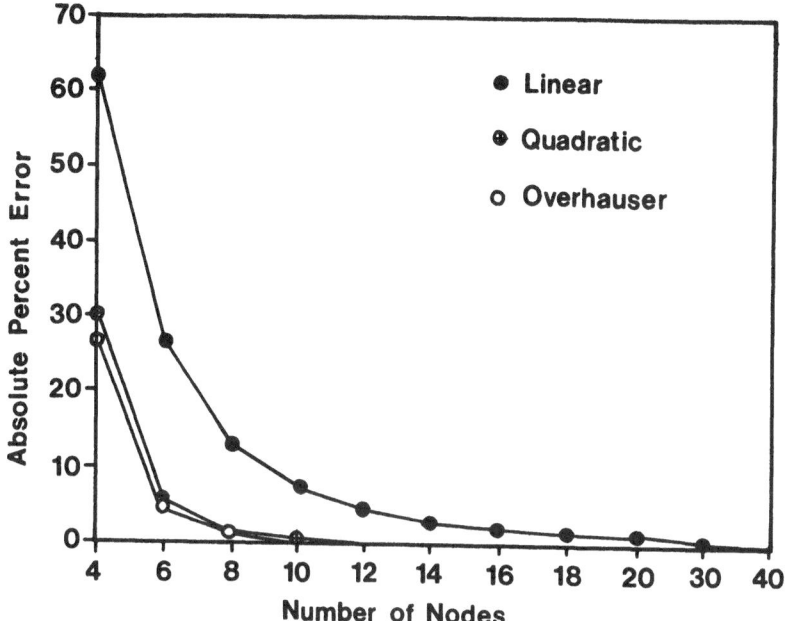

Figure 5.3   Absolute Percent Error For the Center Deflection of a Clamped Circular Plate Under a
Quadratic Load.

TABLE 5.3

Deflection and Moment Function for a Clamped Circular
Plate Under a Quadratic Load $P(r)=q(r/a)^2$.

r/a	α	α exact	β	β exact
0.0	0.03472	0.03472	0.02083	0.02083
0.2	0.03263	0.03263	0.02073	0.02073
0.4	0.02645	0.02646	0.01923	0.01923
0.6	0.01678	0.01678	0.01273	0.01273
0.8	0.00594	0.00594	-0.00476	-0.00476

Note: Deflection, $w=\alpha qa^4(10^{-1})/D$, Moment Function
$M=\beta qa^2$, and radius a.

5.2.1.4 Asymmetric Loading. This example will illustrate the effectiveness of the domain "fanning" technique for a curved geometry. Consider an asymmetric loading of the form $P(r,\theta) = q_0 + q_1(r/a)\cos(\theta)$ acting on a clamped circular plate. The domain integrations involving the loading function cannot be transformed into an equivalent set of surface integrals because of the transcendental function $\cos(\theta)$ in the second term. Deflections of the plate at various internal points are calculated using boundary meshes of 10, 20, and 30 Overhauser elements. The results are compared to an exact solution given by Ugural (1981), and listed in Table 5.4. Previous examples in this section have shown that solutions with a high degree of accuracy are obtainable with minimal boundary discretization when the loading function is "well behaved". When coupled with the domain "fanning" technique the Overhauser formulation can handle more complex loading functions and still retain all of its modelling advantages.

5.2.1.5 Iterative Example - Elastic Foundations. The behavior of structural plates resting on an elastic foundation was first described by Winkler in 1867. In this model the foundation is replaced by an equivalent spring system and applied as an additional loading. Therefore, the governing equation for the deflection of a thin plate on a Winkler foundation is given as

$$\nabla^4 w = \frac{P(x,y)}{D} - \frac{k}{D} w \qquad (5.7)$$

where k is called the modulus of the foundation. The value of the foundation modulus k may vary from 0 to $200MN/m^3$ depending upon the subgrade. A dimensionless form of the foundation modulus was used and is defined as $K = ka^4/D$. As can be seen in Equation (5.7) the unknown deflection function w is not exclusively dependent on the biharmonic operator. Therefore, an iterative solution technique is required.

This particular example demonstrates the combined effect of several of the developments featured in Chapter 4 of this work. The boundary of the circular plate was discretized with 20 Overhauser elements. Since the loading terms are

TABLE 5.4

Deflection of a Clamped Circular Plate Under an Asymmetric
Load $P(r)=q_0+q_1(r/a)\cos(\theta)$.

$\theta=45$	r/a	α	α_{10}	α_{20}	α_{30}
	0.0	0.25000	0.24715	0.24967	0.24993
	0.2	0.24126	0.23799	0.24088	0.24118
	0.4	0.19303	0.18974	0.19266	0.19298
	0.6	0.11688	0.11409	0.11659	0.11688
	0.8	0.03850	0.03682	0.03836	0.03854
$\theta=90$	0.2	0.23040	0.22767	0.23009	0.23033
	0.4	0.17640	0.17404	0.17613	0.17634
	0.6	0.10240	0.10062	0.10219	0.10235
	0.8	0.03240	0.03144	0.03228	0.03237

Note: Deflection, $w=\alpha q_0/D$, $q_0=q_1$, and radius a=2.

TABLE 5.5

Center Deflection for a Clamped Circular Plate Under a Uniform
Load q, on a Winkler Elastic Foundation.

Dimensionless Foundation Modulus, K	α, Results from Ng	α, Results from Costa and Brebbia	α Using 20 Overhauser Elements
0	0.01562	–	0.01562
20	0.01301	0.01279	0.01314
40	0.01112	0.01096	0.01133
60	0.00969	0.00957	0.00989
80	0.00858	0.00846	0.00878
100	0.00768	0.00760	0.00791
120	0.00695	0.00688	0.00717
140	0.00633	0.00628	0.00656
160	0.00581	0.00577	0.00603
180	0.00537	0.00533	0.00558
200	0.00498	0.00495	0.00518

Note: Deflection $w=\alpha qa^4/D$ and radius=a.

functions of the deflection, an iterative solution procedure was used combining the domain "fanning" integration and "solution map" techniques. The results for different values of the dimensionless foundation modulus K, shown in Tables 5.5 and 5.6, are compared with analytical results obtained by Ng (1969) and a boundary element solution presented by Costa and Brebbia (1985). As can be seen in Tables 5.5 and 5.6, accurate results can be obtained using just 20 Overhauser elements. It is interesting to note that the results were conservative when compared to either the analytical or the constant element solution. This type of behavior is not unexpected considering that the initial iteration of the solution is that of a plate without an elastic foundation. When the maximum percent change in the deflection between corresponding solution map points was less than a prescribed value, the solution was considered to have converged. A relaxation factor of 0.5 was found to be effective and solutions were obtained in an average of seven iterations.

5.2.2 Simply Supported Rectangular Plates

Several different rectangular plates under various loading functions will be presented in this section. Unless otherwise specified, the dimensions of each plate will be 0<x<a, 0<y<b. The discontinuity of the surface geometry will require "double noding" of the "extra" node when using Overhauser elements at corners. However, the linear nature of the geometry will allow implementation of the analytical expressions developed in Chapter 3, for the subparametric form of the element. The order of the loading function is increased with each example in an effort to demonstrate the flexibility of the presented formulation. Element performance is evaluated in selected examples so as to be compared with corresponding circular plate analysis.

5.2.2.1 Edge Moments. The bending of a rectangular plate by uniform moments distributed along two parallel sides is considered. This type of loading condition is extremely easy to model. The homogeneous form of the biharmonic equation

TABLE 5.6

Edge Moments for a Clamped Circular Plate Under a Uniform
Load q, on a Winkler Elastic Foundation.

Dimensionless Foundation Modulus, K	α, Results from Ng	α, Results from Costa and Brebbia	α Using 20 Overhauser Elements
0	-0.12500	-	-0.12498
20	-0.10858	-0.10914	-0.10875
40	-0.09666	-0.09746	-0.09700
60	-0.08760	-0.08851	-0.08765
80	-0.08047	-0.08144	-0.08040
100	-0.07470	-0.07471	-0.07450
120	-0.06993	-0.07095	-0.06960
140	-0.06592	-0.06694	-0.06545
160	-0.06249	-0.06352	-0.06188
180	-0.05953	-0.06054	-0.05993
200	-0.05694	-0.05760	-0.05722

Note: Deflection $w = \alpha a^2 q$ and radius=a.

TABLE 5.7

Center Deflection and Moment Function for a Simply
Supported Rectangular Plate Bent by Moments
Distributed Along Two Parallel Edges.

b/a	α	α exact	β	β exact
0.5	0.0966	0.0964	0.8912	0.8900
1.0	0.0369	0.0368	0.5009	0.5000
1.5	0.0281	0.0280	0.2390	0.2385
2.0	0.0174	0.0174	0.1100	0.1100

Note: Deflection, $w = \alpha M_0 b^2 / D$ for b/a<1,
$w = \alpha M_0 a^2 / D$ for b/a>1, and Moment
Function $M = \beta M_0$.

is solved with the deflection specified at zero along each edge. The Laplacian of w, $\nabla^2 w$ is specified as zero at $x = 0$ and $x = a$, and set equal to $-M_n/D$ at $y = 0$ and $y = b$. Results for the deflection $w(a/2,b/2)$ and the moment function $M(a/2,b/2)$ at several ratios of b/a, shown in Table 5.7, compare very well with analytical results obtained by Timoshenko and Woinowsky-Krieger (1959). The number of Overhauser elements used for this example ranged from 20 for $b/a = 0.5$ to 40 for $b/a = 2.0$ maintaining approximately the same element length throughout the analysis.

5.2.2.2 <u>Thermal Loads</u>. Consider the special case of a simply supported rectangular plate bent by uniform edge moments which are caused by a temperature variation in the plate. Assume the upper surface of the plate is held at a different temperature than the lower surface. The resulting form of the normal edge moment for the linear temperature distribution is given as

$$M_n = \frac{\alpha t (1+\nu)}{h} \tag{5.8}$$

where t is the temperature difference between the upper and lower surfaces, α is the coefficient of thermal expansion, h is the plate thickness, and ν is Poisson's ratio. The center deflection of the plate for several ratios of a/b are compared to an analytical solution given by Timoshenko and Woinowsky-Krieger (1959) and listed in Table 5.8. The modelling procedure was identical to that of the previous example.

5.2.2.3 <u>Concentrated Load</u>. This example will demonstrate the accuracy of the boundary element solution for concentrated loads over rectangular geometries. As previously discussed in the circular plate example, a concentrated load may be modelled quite easily. The reduction of each domain integral involving the load to a single evaluation combined with the analytic expressions available for linear geometries provides an excellent foundation for a very accurate numerical solution. The results for the center deflection $w(a/2,b/2)$ for various ratios of

TABLE 5.8

Center Deflection for a Simply Supported
Rectangular Plate Bent by Thermal Loads.

a/b	α	α exact
0.5	0.02846	0.02847
1.0	0.07367	0.07367
1.5	0.10074	0.10077
2.0	0.11385	0.11387
5.0	0.12489	0.12490
10.0	0.12490	0.12500

Note: Deflection, $w = \alpha t(1+\nu)a^4/h$

TABLE 5.9

Center Deflection for a Simply Supported
Rectangular Plate Bent by a Concentrated
Load, P, Located at its Center.

b/a	α	α exact
1.0	0.01160	0.01160
1.2	0.01355	0.01354
1.4	0.01486	0.01484
1.6	0.01570	0.01570
1.8	0.01621	0.01620
2.0	0.01652	0.01651

Note: Deflection, $w = \alpha Pa^2/D$.

b/a are given in Table 5.9. Solutions were obtained using Overhauser elements coupled with a spacing scheme similar to the one used in the preceding example.

The next four examples incrementally increase the order of the loading function. Each domain term representing the loading function is transformed into an equivalent series of surface integrals. Higher order functions for the load naturally required more surface integrations to evaluate the effects. The proceeding analysis demonstrates the effectiveness of the integral transformations.

5.2.2.4 <u>Uniform Load.</u> The behavior of a uniformly loaded square plate is presented in this example. For a linear element formulation, the domain integrals representing the effects of the transverse load are evaluated using the analytical expressions derived in Chapters 3 and 4. For quadratic and Overhauser analyses, the surface integral terms are also evaluated from analytical expressions developed previously. However, the transformed surface integral representing the load effects and the "corner" Overhauser element are calculated using numerical quadrature.

Absolute percentage error for the center deflection between the results obtained using each element type and the analytical solution given by Szilard (1974) is shown in Figure 5.4 for various boundary mesh sizes. The Overhauser element formulation was determined to be superior to the other two types of elements. However, it should be noted that a quadratic solution using 18 nodes had a lower absolute percent error than an equivalent Overhauser analysis. For meshes of 24 nodes and above, the solutions given by either of the two elements were indistinguishable. The absolute percent error for this solution over a linear geometry when compared at equivalent sized meshes is lower then the circular geometry presented earlier in Figure 5.2. This result is not unexpected considering the ability of each element type to exactly represent linear geometries. Furthermore, the analytic expressions developed in Chapter 3 for the surface integrations over linear geometries provide additional accuracy in the approximation.

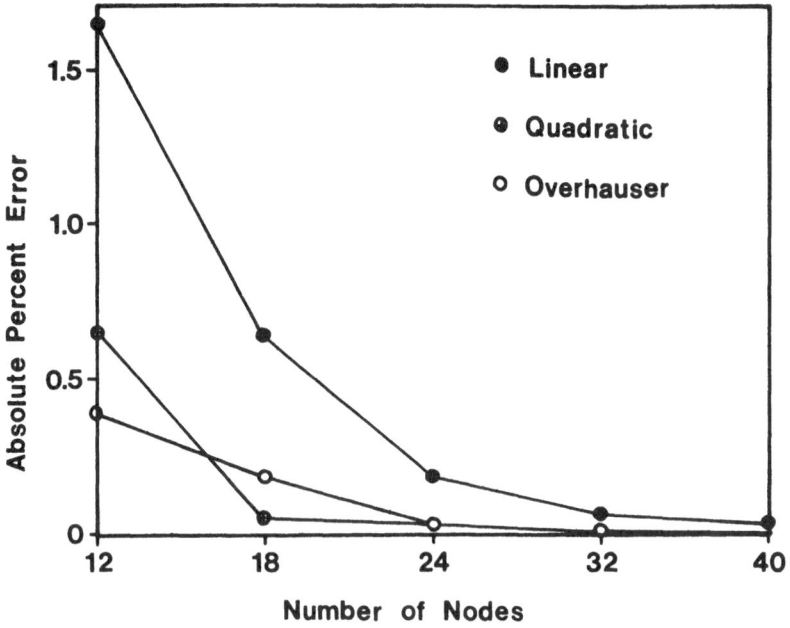

Figure 5.4 Absolute Percent Error For the Center Deflection of a Simply Supported Square Plate
Under a Uniform Load.

The solution presented here will also serve to verify the accuracy of the deflection and its first derivative, as well as the moment function and its derivatives. The results using 24 Overhasuer elements are compared with analytical solutions given by Ugural (1981) and listed in Table 5.10. The accuracy of the derivatives will be important in evaluating more complex domain terms presented in later examples.

5.2.2.5 Hydrostatic Load. Consider a simply supported square plate loaded by a hydrostatic load of the form q(x/a). The loading function is transformed using the harmonic form of the integral transformation described in Chapter 4. Results for the deflection, the moment function, and their derivatives obtained using 24 Overhauser elements are given in Table 5.11. The domain integrals for this type of loading are replaced by a set of two surface integrals, each of which are evaluated numerically.

5.2.2.6 Quadratic Load. A simply supported square plate under a quadratic load of the form $q(x/a)^2$ is presented in the third example as an illustration of the integral transformation technique. Each domain integral representing the loading function is transformed into a series of three surface integrals. Results for the deflection, the moment function, and their derivatives using a discretization identical to that presented in the preceding two examples are listed in Table 5.12. Absolute percent error between this solution and the analytical results given by Timoshenko and Woinowsky-Krieger (1959) is shown in Figure 5.5 for several mesh sizes. The scale of the error is smaller than that in the corresponding circular plate examples with the same order loading functions. This is due to the fact that when modelling linear geometries, the integrations for all element types are exact. For a small number of elements the accuracy of the Overhauser formulation suffers from the influence of the "corner" element. However, as more nodes are used to describe the boundary the "corner" effect is

TABLE 5.10

Deflection, Moment Function, and Their Derivatives at Various
Points on a Uniformly Loaded Square Plate.

(x,y)	(a/2,b/4)	(a/2,b/2)	(3a/4,b/4)	(3a/4,b/2)
α	0.00295	0.00406	0.00213	0.00294
α exact	0.00294	0.00406	0.00213	0.00294
α_1	0.00000	0.00000	-0.00631	-0.00877
α_1 exact	0.00000	0.00000	-0.00630	-0.00876
α_2	0.00875	0.00001	0.00629	0.00000
α_2 exact	0.00876	0.00000	0.00630	0.00000
β	0.05734	0.07368	0.04531	0.05734
β exact	0.05733	0.07367	0.04529	0.05733
β_1	0.00000	0.00000	-0.10189	-0.13639
β_1 exact	0.00000	0.00000	-0.10196	-0.13637
$\beta2$	0.13639	0.00000	0.10196	0.00000
β_2 exact	0.13637	0.00000	0.10196	0.00000

Note: Deflection $w=\alpha qa^4/D$, $\partial w/\partial x_i=\alpha_i qa^3/D$, Moment Function
$M=\beta qa^2$, and $\partial M/\partial x_i=\beta_i qa$.

122

TABLE 5.11

Deflection, Moment Function, and Their Derivatives at Various
Points on a Square Plate Under a Hydrostatic Load.

(x,y)	(a/2,b/4)	(a/2,b/2)	(3a/4,b/4)	(3a/4,b/2)
α	0.00147	0.00203	0.00119	0.00163
α exact	0.00147	0.00203	0.00119	0.00163
α_1	0.00071	0.00093	-0.00309	-0.00431
α_1 exact	0.00071	0.00093	-0.00308	-0.00431
α_2	0.00437	0.00001	0.00345	0.00001
α_2 exact	0.00438	0.00000	0.00346	0.00000
β	0.02867	0.03684	0.02873	0.03579
β exact	0.02867	0.03684	0.02871	0.03578
β_1	0.03122	0.03735	-0.04103	-0.05782
β_1 exact	0.03104	0.03718	-0.04146	-0.05817
β_2	0.06819	0.00000	0.06091	0.00000
β_2 exact	0.06824	0.00000	0.06107	0.00000

Note: Deflection $w=\alpha qa^4/D$, $\partial w/Mx_i=\alpha_i qa^3/D$, Moment Function
$M=\beta qa^2$, and $\partial M/\partial x_i=\beta_i qa$.

TABLE 5.12

Deflection, Moment Function, and Their Derivatives at Various
Points on a Square Plate Under a Quadratic Load.

(x,y)	(a/2,b/4)	(a/2,b/2)	(3a/4,b/4)	(3a/4,b/2)
α	0.00867	0.01200	0.00761	0.01038
α exact	0.00867	0.01200	0.00761	0.01038
α_1	0.00706	0.00927	-0.01756	-0.02471
α_1 exact	0.00711	0.00933	-0.01750	-0.02463
α_2	0.02585	0.00006	0.02181	0.00006
α_2 exact	0.02591	0.00000	0.02191	0.00000
β	0.01642	0.02121	0.01994	0.02460
β exact	0.01642	0.02121	0.01992	0.02460
β_1	0.03122	0.03735	-0.01433	-0.02366
β_1 exact	0.03119	0.03733	-0.01439	-0.02364
β_2	0.03985	0.00000	0.04075	0.00000
β_2 exact	0.03984	0.00000	0.04082	0.00000

Note: Deflection $w=\alpha qa^4(10^{-1})/D$, $\partial w/\partial x_i = \alpha_i qa^3(10^{-1})/D$, Moment
Function $M=\beta qa^2$, and $\partial M/\partial x_i = \beta_i qa$.

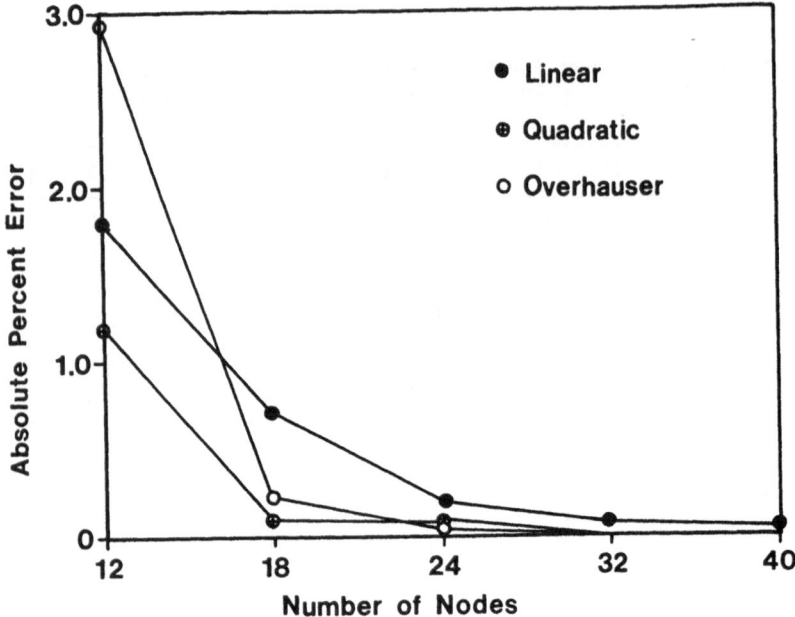

Figure 5.5 Absolute Percent Error For the Center Deflection of a Simply Supported Square Plate
Under a Quadratic Load.

negligible and the Overhauser again demonstrates its superiority over the other two elements.

5.2.2.7 <u>Cubic Loading</u>. The final example illustrating the accuracy of the integral transformation technique will be that of a simply supported square plate with a transverse loading function of the form $q(x/a)^3$. Both of the domain integrals representing the effects of the cubic loading function are converted into a series of four surface integrations using the biharmonic form of the transformation. This example demonstrates a limiting case of the integral transformation technique for the formulation presented in this work. Any loading function of a higher order will be integrated using the domain "fanning" quadrature scheme. Results for the deflection, moment function, and their derivatives using a 24 element Overhauser formulation are given in Table 5.13.

The results of the four preceding analyses listed in Tables 5.10, 5.11, 5.12, and 5.13 are in excellent agreement with existing analytic solutions. Calculations for the derivatives of the deflection and the moment function are determined to be very accurate. This analysis is important in validating the ability of the formulation presented in Chapter 3 and 4 in calculating the derivatives of the field variables ψ and ω. The next series of example problems involve complex "loading" functions for which accurate values of the field variables and their derivatives are required to evaluate the necessary domain integrals.

5.2.2.8 <u>Iterative Example - Elastic Foundations</u>. Consider the behavior of a uniformly loaded simply supported square plate resting on an elastic foundation. The governing equation defined in Equation (5.7) has the unknown deflection as part of the "loading" function. An iterative solution using both the domain "fanning" and solution map techniques was used to solve this problem. The center deflection for various values of the K, the dimensionless foundation modulus, are compared to analytical results given by Ugural (1981) in Table 5.14. The number of iterations necessary to meet specified convergence criteria varied from 4 to

TABLE 5.13

Deflection, Moment Function, and Their Derivatives at Various
Points on a Square Plate Under a Cubic Load.

(x,y)	(a/2,b/4)	(a/2,b/2)	(3a/4,b/4)	(3a/4,b/2)
α	0.00566	0.00784	0.00529	0.00719
α exact	0.00565	0.00784	0.00528	0.00719
α_1	0.00591	0.00777	-0.01091	-0.01552
α_1 exact	0.00597	0.00783	-0.01086	-0.01536
α_2	0.01691	0.00007	0.01503	0.00007
α_2 exact	0.01697	0.00000	0.01510	0.00000
β	0.01029	0.01340	0.01459	0.01790
β exact	0.01029	0.01340	0.01457	0.01789
β_1	0.02557	0.03072	-0.00142	-0.00703
β_1 exact	0.02555	0.03070	-0.00147	-0.00702
β_2	0.02568	0.00000	0.02914	0.00000
β_2 exact	0.02567	0.00000	0.02920	0.00000

Note: Deflection $w=\alpha qa^4(10^{-1})/D$, $\partial w/\partial x_i=\alpha_i qa^3(10^{-1})/D$, Moment
Function $M=\beta qa^2$, and $\partial M/\partial x_i=\beta_i qa$.

TABLE 5.14

Center Deflection for a Simply Supported Plate
Under a Uniform Load q, on a Winkler
Elastic Foundation

Dimensionless Foundation Modulus, K	α, Exact Results	α Using 24 Overhauser Elements
0	0.04062	0.04064
16	0.03898	0.03904
32	0.03747	0.03759
48	0.03607	0.03614
64	0.03476	0.03482
80	0.03354	0.03371
160	0.02853	0.02888
240	0.02479	0.02484

Note: Deflection $w = \alpha a^4 q (10^{-1})/D$.

TABLE 5.15

Center Deflection for a Simply Supported Plate
Under a Hydrostatic Load q(x/a), on a
Winkler Elastic Foundation.

Dimensionless Foundation Modulus, K	α, Exact Results	α Using 24 Overhauser Elements
0	0.02031	0.02032
16	0.01949	0.01956
32	0.01873	0.01881
48	0.01803	0.01810
64	0.01738	0.01746
80	0.01677	0.01687
160	0.01426	0.01432
240	0.01240	0.01245

Note: Deflection $w = \alpha a^4 q (10^{-1}/D$.

15 for the values of K equal to 16 and 240 respectively. The analysis was extended to include a hydrostatic loading of the form q(x/a) and a quadratic load
given as q(xy/a^2). Results obtained for these cases are compared to analytical
expressions and presented in Tables 5.15 and 5.16. The increase in order of the
loading functions had little or no effect on the number of iterations required
for convergence.

5.2.2.9 Iterative Example - In-Plane Forces. Consider the flexural behavior of
a simply supported rectangular plate under the combined action of a uniform lateral load and uniform in-plane force. The governing equation for the deflection
is defined as

$$\nabla^4 w = \frac{1}{D} \left(q + N_x \frac{\partial^2 w}{\partial x^2} + N_y \frac{\partial^2 w}{\partial y^2} + 2N_{xy} \frac{\partial^2 w}{\partial x \partial y} \right) \tag{5.9}$$

where N_x and N_y are normal forces in the x and y directions respectively and N_{xy}
is the shearing force. If N_x and N_y are equal to N_f and N_{xy} is zero, the
governing equation reduces to

$$\nabla^4 w = \frac{q}{D} \left(1 + N \nabla^2 w \right) \tag{5.10}$$

where N is a parameter defined as N_f/q. The right-hand side of the equation
contains the term $\nabla^2 w$ and requires an iterative solution procedure when solved by
the technique presented in Chapter 2. Results for various values of N for several ratios of a/b are compared to an analytical solution given by Timoshenko and
Woinowsky-Krieger, (1959), and presented in Table 5.17. At large negative values
of the parameter N, the numerical solution experienced difficulty in converging.
In fact, the last case where N = -3 and the ratio a/b = 2.0 the iterative solution technique diverged. These results were not completely unexpected, since for
values of N_x = 4.46D, N_y = 0, and no transverse loading, a square plate reaches
its first buckling mode.

5.2.2.10 Iterative Example - Variable Thickness. A simply supported square
plate of variable thickness is considered. Assuming no discontinuous changes in

TABLE 5.16

Center Deflection for a Simply Supported Plate Under
a Quadratic Load $q(xy/a^2)$, on a Winkler Elastic
Foundation. Deflection $w=\alpha a^4 q(10^{-2})/D$.

Dimensionless Foundation Modulus, K	α, Exact Results	α Using 24 Overhauser Elements
0	0.10156	0.10160
16	0.09746	0.09757
32	0.09367	0.09381
48	0.09016	0.09030
64	0.08690	0.08713
80	0.08382	0.08417
160	0.07132	0.07153
240	0.06198	0.06220

TABLE 5.17

Center Deflection for a Simply Supported Rectangular Plate
Under the Combined Action of Uniform Lateral and Uniform
In-Plane Forces. Deflection $w=\alpha q a^4/D$.

a/b	0.5		1.0	
N	α	α exact	α	α exact
3	0.000508	0.000506	0.002501	0.002501
2	0.000541	0.000542	0.002895	0.002870
1	0.000586	0.000584	0.003376	0.003365
0	0.000633	0.000633	0.004062	0.004062
-1	0.000689	0.000691	0.005083	0.005115
-2	0.000757	0.000760	0.006785	0.006888
-3	0.000839	0.000844	0.010163	0.010499
a/b	1.5		2.0	
3	0.004197	0.004126	0.005024	0.005018
2	0.004879	0.004891	0.006140	0.006042
1	0.006036	0.005994	0.007584	0.007633
0	0.007724	0.007724	0.010129	0.010129
-1	0.010731	0.010814	0.014973	0.015165
-2	0.017377	0.017878	0.028447	0.029603
-3	0.044809	0.050074	-	-

the thickness, the governing equation for bending is given as

$$D\nabla^4 w + 2 \frac{\partial D}{\partial x} \frac{\partial}{\partial x} \nabla^2 w + 2 \frac{\partial D}{\partial y} \frac{\partial}{\partial y} \nabla^2 w - \nabla^2 D \nabla^2 w$$

$$- (1-\nu)(\frac{\partial^2 D}{\partial x^2} \frac{\partial^2 w}{\partial y^2} - 2 \frac{\partial^2 D}{\partial x \partial y} \frac{\partial^2 w}{\partial x \partial y}$$

$$+ \frac{\partial^2 D}{\partial y^2} \frac{\partial^2 w}{\partial x^2}) = P(x,y)$$

(5.11)

The flexural rigidity is no longer a constant. For this example, it was considered to be a function of y only and is given as $D = D_0 + D_1 y$. The relationship
between D_0 and D_1 for this example was $D_1 = 7D_0/b$. Equation (5.11) reduces to
the following form

$$\nabla^4 w = \frac{q_0}{D_0} (1 - \frac{2 (\frac{\partial \nabla^2 w}{\partial y})}{q_0 (1 + 7 \frac{y}{b})})$$

(5.12)

The deflection and the moment function at points along x = a/2, shown in Figures
5.6 and 5.7 respectively, are compared to numerical results presented in Timoshenko and Woinowsky-Krieger (1959). The numerical solution given by Timoshenko
is very difficult to obtain, whereas the boundary element formulation developed
in this work provided accurate results in only four iterations.

5.2.3 Rectangular Plates with Various Edge Conditions

5.2.3.1 Introduction

A series of examples combining simple and clamped edge conditions for rectangular plates will be presented in the following section. In each case the
boundary was described by Overhauser elements using analytical expressions for
the necessary surface integrations. At points where the boundary condition
abruptly changes from a simple support to a clamped edge, "double" noding was

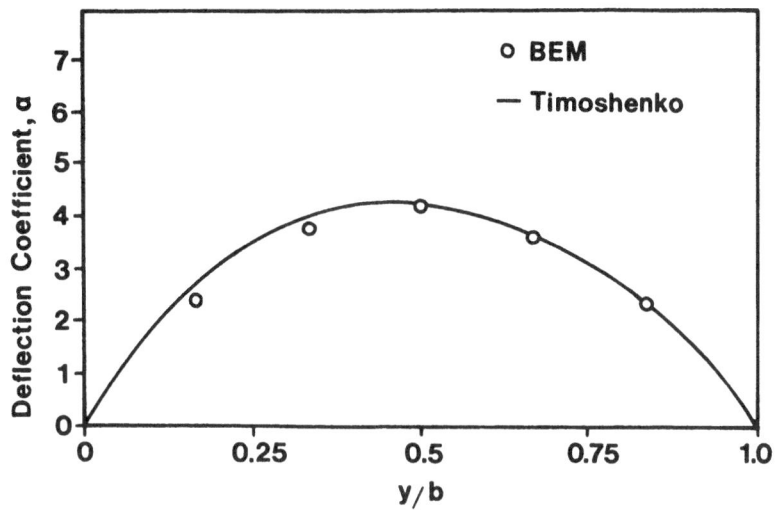

Figure 5.6   Deflection Along the Centerline, x=a/2, for a
Simply Supported Plate of Variable Thickness.
Deflection $w=\alpha q_0 a^4 (10^{-3})/D_0$.

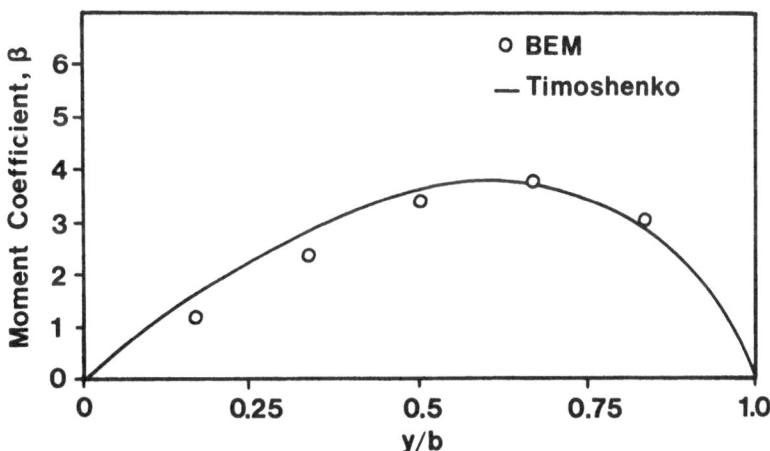

Figure 5.7   Moment Function Along the Centerline, x=a/2,
for a Simply Supported Plate of Variable
Thickness.   Moment Function $M=\beta q_0 a^2 (10^{-2})$.

used to accurately model its effects. Numerical quadrature was used to evaluate any integrations over the corner version of the Overhauser element and the transformed domain integrals involving the loading function. Results for each example are compared to analytical solutions given by Timoshenko and Woinowsky-Krieger (1959) and are presented in tabular form.

Conspicuously absent in this example section are cases in which iteration is required. The reason for this is the lack of published examples for which comparisons can be made. However, the program is quite capable of handling these cases.

5.2.3.2 <u>One Clamped Edge</u>. The flexural response of a rectangular plate with three simply supported edges and one clamped edge (at y = 0) are presented here. Values for the deflection and moment function at the center of the plate for various ratios of b/a for both a uniform and a hydrostatic loading are given in Tables 5.18 and 5.19 respectively.

5.2.3.3 <u>Two Opposite Edges Clamped</u>. Consider a rectangular plate where two opposite edges are simply supported and the other two edges are clamped. Two loading cases were examined: a uniform load q, and a hydrostatic of the form q(x/a). For both cases each corner is "double" noded to handle the discontinuous boundary conditions. A boundary element solution using an Overhauser formulation for both loading cases is compared to a corresponding analytical solution and listed in Tables 5.20 and 5.21.

5.2.3.4 <u>All Edges Clamped</u>. In this example the deflection of a rectangular plate with all edges clamped is presented. As with the preceding examples, two loading cases were examined: a uniform load q, and a hydrostatic load q(x/a). The deflection and the moment function at the center of the plate is calculated for several ratios of b/a and compared to exact solutions in Tables 5.22 and 5.23.

TABLE 5.18

Center Deflection and Moment Function for a
Simply Supported Rectangular Plate with One
Edge Clamped Bent by a Uniform Load q.

b/a	α	α exact	β	β exact
0.5	0.0049	0.0049	0.0648	0.0638
1.0	0.0028	0.0028	0.0562	0.0561
1.5	0.0064	0.0064	0.0899	0.0900
2.0	0.0093	0.0093	0.1084	0.1085

Note: Deflection, $w=\alpha qb^4/D$ for b/a<1, $w=\alpha qa^4/D$
       for b/a>1, Moment Function $M=\beta qb^2$ for
       b/a<1, and $M=\beta qa^2$ for b/a>1.

TABLE 5.19

Center Deflection and Moment Function for a
Simply Supported Rectangular Plate with
One Edge Clamped Bent by a Hydrostatic
Load q(x/a).

b/a	α	α exact	β	β exact
0.5	0.0045	0.0045	0.0533	0.0538
1.0	0.0013	0.0013	0.0266	0.0269
1.5	0.0019	0.0019	0.0298	0.0300
2.0	0.0022	0.0023	0.0300	0.0308

Note: Deflection, $w=\alpha qb^4/D$ for b/a<1, $w=\alpha qa^4/D$
       for b/a>1, Moment Function $M=\beta qb^2$ for
       b/a<1, and $M=\beta qa^2$ for b/a>1.

TABLE 5.20

Center Deflection and Moment Function for a Rectangular
Plate with Two Opposite Edges Clamped and the Other
Two Simply Supported Bent by a Uniform Load q.

b/a	α	α exact	β	β exact
0.5	0.00257	0.00260	0.04291	0.04321
1.0	0.00191	0.00192	0.04423	0.04431
1.5	0.00532	0.00531	0.08029	0.08038
2.0	0.00844	0.00844	0.10323	0.10331

Note: Deflection, $w=\alpha qb^4/D$ for b/a<1, $w=\alpha qa^4/D$ for
b/a>1, Moment Function $M=\beta qb^2$ for b/a<1, and
$M=\beta qa^2$ for b/a>1.

TABLE 5.21

Center Deflection and Moment Function for a
Rectangular Plate with Two Opposite Edges
Clamped and the Other Two Simply Sup-
ported Bent by a Hydrostatic
Load q(x/a).

b/a	α	β	β exact
0.5	0.00128	0.02146	0.02154
1.0	0.00096	0.02212	0.02308
1.5	0.00266	0.04015	0.04077
2.0	0.00422	0.05164	0.05154

Note: Deflection, $w=\alpha qb^4/D$ for b/a<1, $w=\alpha qa^4/D$
for b/a>1, Moment Function $M=\beta qb^2$ for
b/a<1, and $M=\beta qa^2$ for b/a>1.

TABLE 5.22

Center Deflection and Moment Function for a
Rectangular Plate with All Edges Clamped
Bent by a Uniform Load q.

b/a	α	α exact	β	β exact
0.5	0.00016	0.00016	0.00109	0.00109
1.0	0.00127	0.00126	0.03526	0.03554
1.5	0.00220	0.00220	0.04388	0.04392
2.0	0.00253	0.00254	0.04382	0.04385

Note: Deflection, $w=\alpha qa^4/D$ and Moment Function $M=\beta qa^2$.

TABLE 5.23

Center Deflection and Moment Function for a
Rectangular Plate with All Edges Clamped
Bent by a Hydrostatic Load q.

b/a	α	α exact	β	β exact
0.5	0.00008	0.00008	0.00548	0.00548
1.0	0.00063	0.00063	0.01763	0.01769
1.5	0.00110	0.00110	0.02194	0.02200
2.0	0.00127	0.00128	0.02191	0.02192

Note: Deflection, $w=\alpha qa^4/D$ and Moment Function $M=\beta qa^2$.

TABLE 5.24

Center Deflection for a Simply Supported
Skewed Plate Bent by a Uniform Load q.

θ	m	α exact	α
0	2.00	0.01013	0.01013
30	2.02	0.01046	0.00989
45	2.00	0.00938	0.00895
60	2.00	0.00796	0.00653
75	2.00	0.00094	0.00097

Note: Deflection $w=\alpha qa^4/D$.

5.2.3.5 <u>Iterative Example - All Edges Clamped, Elastic Foundation.</u> A solution for the deflection and moment function of a uniformly loaded rectangular plate with clamped edges on a Winkler type elastic foundation is presented. The governing equation for the deflection, given by Equation (5.7), was solved using an iterative solution technique identical to that used for the preceding elastic foundation problems. Results for the center deflection and maximum value of the moment function at the edge for a dimensionless foundation modulus of K = 200 (ka^4/D) for various aspect ratios are shown in Figures 5.8 and 5.9. These results are in excellent agreement with numerical solutions of Costa and Brebbia (1985), and results using a Galerkin variational method given by Ng (1969).

5.2.4 <u>Plates of Various Shapes</u>

5.2.4.1 <u>Introduction</u>

In this example several polygonal shaped plates with both simply and clamped edge supports for various loading functions are presented. In each case the boundary element formulation with the Overhauser element developed in Chapters 3 and 4 of this work was used to obtain solutions for the deflection and the moment function. Results are compared to analytical or published numerical solutions to verify their accuracy.

5.2.4.2 <u>Simply Supported Triangular Plates.</u> Consider a simply supported equilateral triangular plate under two loading conditions: a uniformly distributed moment M_n applied along the boundary and a uniform load q. The deflection along a line, of length a, that bisects one side and passes through the opposite vertex for each loading condition is shown in Figures 5.10 and 5.11 and compared to an analytical solution given by Timoshenko and Woinowsky-Krieger (1959).

5.2.4.3 <u>Triangular Plate with Two or Three Edges Clamped.</u> In this example, the deflection of an equilateral triangular plate along the centerline defined in the previous problem for both a uniform load q and hydrostatic load q(x/a) are pre-

137

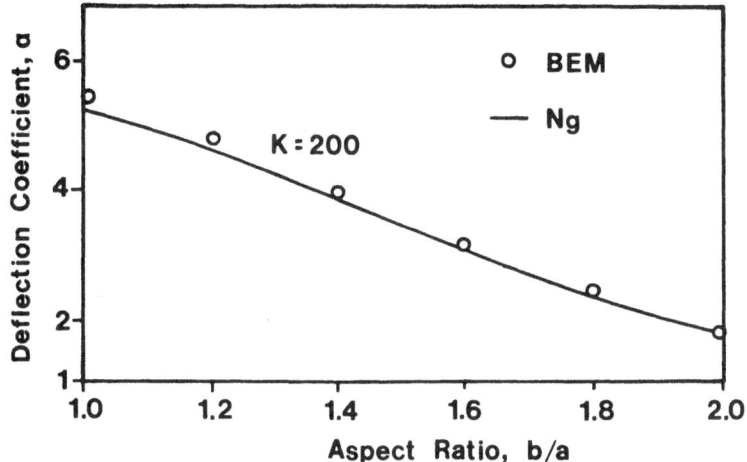

Figure 5.8   Center Deflection for a Clamped Rectangular
Plate on an Elastic Foundation. Deflec-
tion w=αqb⁴(10⁻³)/D.

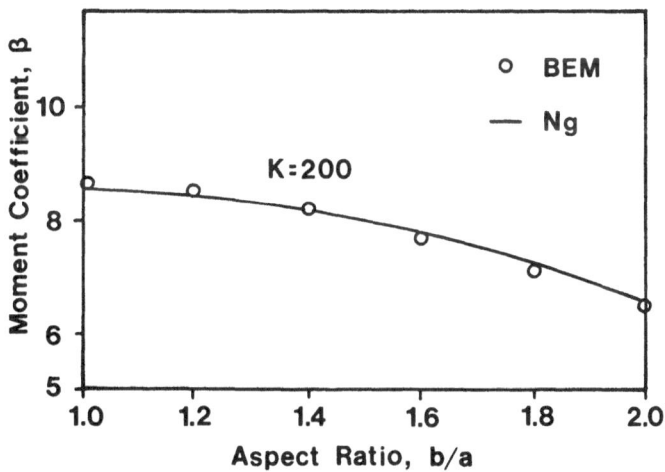

Figure 5.9   Maximum Edge Moment for a Clamped Rectangular Plate on an Elastic Foundation.
Edge Moment M=βqb²(10⁻²).

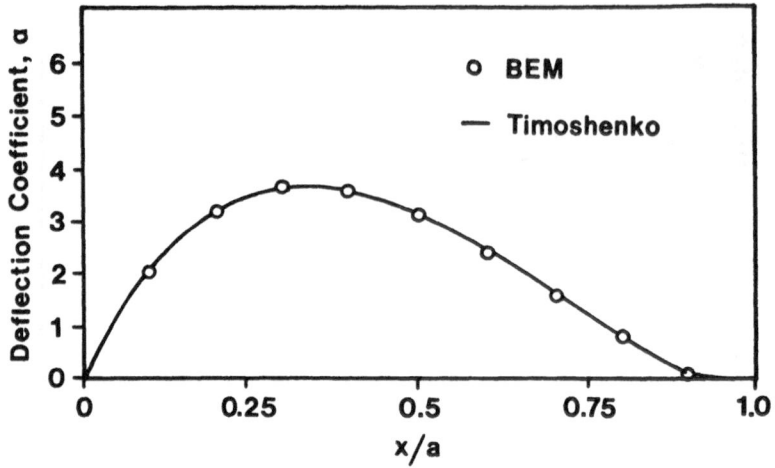

Figure 5.10   Deflection Along the Centerline of a Simply
Supported Triangular Plate Bent by Uniform
Edge Moments.   Deflection $w=\alpha M_n a^4 (10^{-2})/D$.

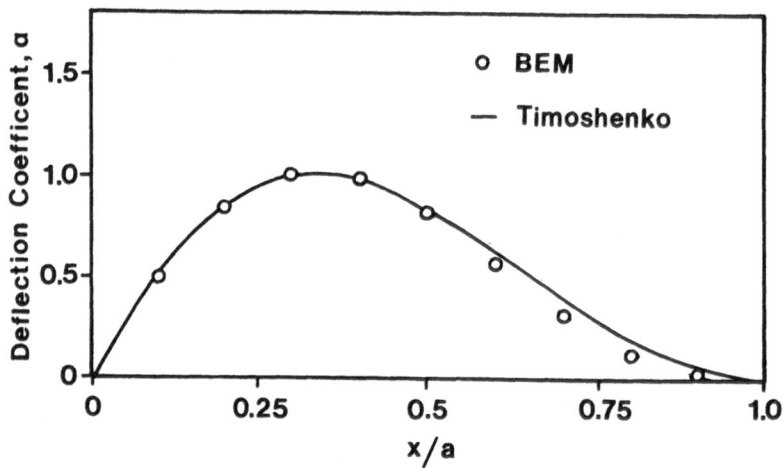

Figure 5.11   Deflection Along the Centerline of a Simply
Supported Triangular Plate Bent by a Uni-
form Load.   Deflection $w=\alpha q a^4 (10^{-3})/D$.

sented. In Figure 5.12, the deflection for a plate where the two sides are clamped while the remaining side is simply supported is shown for both loading functions. Results for the deflection of an equilateral triangular plate where all edges are clamped are presented in Figure 5.13. Corners where the edge conditions changed from clamped to simple supports were modelled effectively by using the "double" noding technique describe earlier.

5.2.4.4 <u>Skewed Plates</u>. In this example the deflection at the center of a simply supported oblique parallelogram shaped plate, shown in Figure 5.14, is presented. This type of plate could be used as a floor slab in skewed bridge. Results for various angles θ are compared with numerical solutions given by Timoshenko and Woinowsky-Krieger (1959) and listed in Table 5.24.

5.2.4.5 <u>Rhombic Plates</u>. This case will examine the flexural behavior of a simply supported rhombic plate shown in Figure 5.15(a). Results for the deflection and bending moment at the center of the plate for various values of the angle θ are listed in Table 5.25. It can be seen that the results using 12 Overhauser elements are in excellent agreement will those of Maiti and Chakrabarty (1974) using 32 constant elements or Leissa (1965) obtained by a variational approach.

5.2.4.6 <u>Hexagonal Plates</u>. A uniformly loaded simply supported hexagonal plate, shown in Figure 5.15(b), is considered in this example. Results for several mesh sizes are compared to a published numerical solution given by Maiti and Chakrabarty (1974) and a solution obtained by Leissa (1965) using a variational method. A comparison of the values of the deflection and the moment function at the center of the hexagonal plate, presented in Table 5.26, indicates the formulation developed in this work is in excellent agreement with existing solutions.

5.2.4.7 <u>Corner Plate</u>. Corner plates are used to analyze polygonal shaped plates with a polygonal cut-out. Triangular and many other polygonal shaped plates are

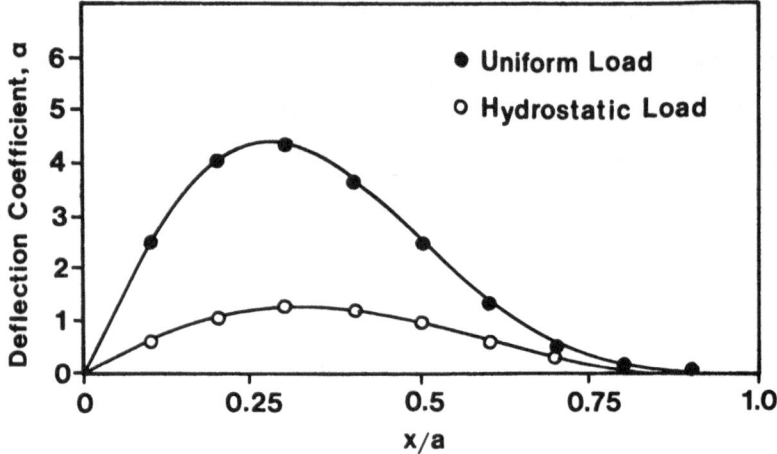

Figure 5.12 Deflection Along the Centerline of a Triangular Plate with Two Sides Clamped.
Deflection $w = \alpha q a^4 (10^{-4})/D$.

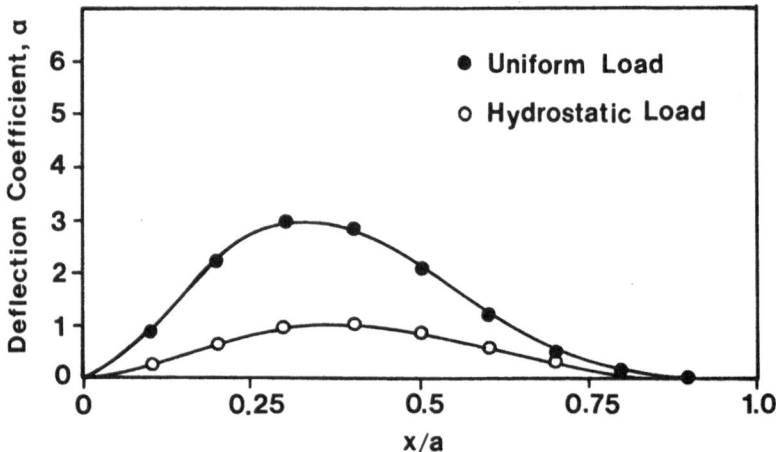

Figure 5.13 Deflection Along the Centerline of a Triangular Plate with All Edges Clamped.
Deflection $w = \alpha q a^4 (10^{-4})/D$.

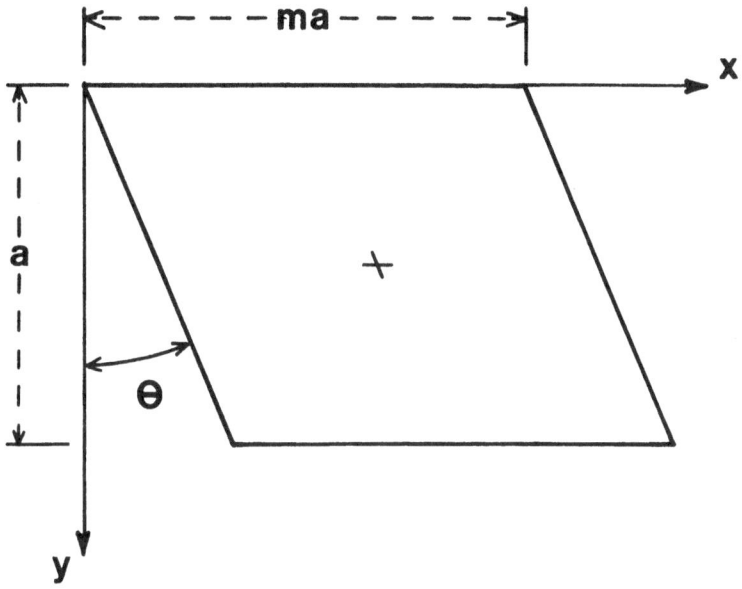

Figure 5.14   Skewed Plate Geometry.

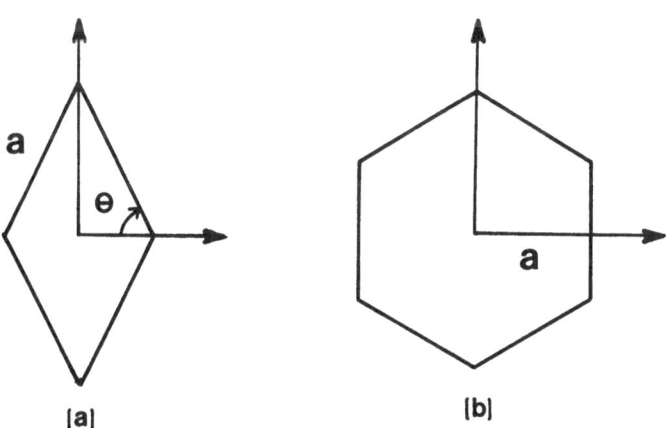

(a)   (b)

Figure 5.15   Rhombic and Hexagonal Plate Geometries.

TABLE 5.25

Center Deflection and Moment Function for a Simply Supported
Rhombic Plate Under a Uniform Load q.

	Angle θ	Using 24 Overhasuer Elements	Using 32 Constant Elements	Variational Approach
α	45	0.00408	0.00408	-
	60	0.00256	0.00256	0.00256
	75	0.00041	0.00038	0.00041
β	45	0.07387	0.07385	-
	60	0.05838	0.05838	0.05831
	75	0.02315	0.02292	0.02300

Note: Deflection $w=\alpha qa^4/D$ and Moment Function $M=\beta qa^2$.

TABLE 5.26

Center Deflection and Moment Function for a Simply
Supported Hexagonal Plate Under a Uniform Load q.

Number of Nodes	Results using Overhauser Formulation		Results from Maiti and Chakrabarty (1974)	
	α	β	α	β
12	0.0573	0.27256	0.0550	0.27077
24	0.0548	0.27019	0.0547	0.27023
48	0.0546	0.26989	-	-
Results of Leissa (1965)	0.0548	0.27077	0.0548	0.27077

Note: Deflection $w=\alpha qa^4/D$ and Moment Function $M=\beta qa^2$.

defined by the angle between linear segments as shown in Figure 5.16. By invoking symmetry, only the corner section of each limb is modelled, see Figure 5.17. At lines of symmetry the normal derivatives of both the deflection, $\partial w/\partial n$, and the Laplacian of the deflection, $\partial(\nabla^2 w)/\partial n$, are set equal to zero. The rest of the boundary is simply supported. Values of the deflection and the moment function along the diagonal, for different angles, are shown in Figures 5.18 and 5.19. The results compare with good accuracy to those presented by Segedin and Brickell (1968).

5.3 Incompressible Viscous Fluid Flow at Low Reynolds Numbers

5.3.1 Introduction

Four examples are presented in this section for the purpose of demonstrating the versatility of this formulation. The first example is a study of the flow field generated by inflow-outflow in a cylinder. The second case is a moving-wall problem in which the domain is completely enclosed. These two problems were originally analyzed by Mills (1977). The third case will examine fluid flow through an array of impermeable cylindrical fibers. This same example was presented by Hildyard, et al (1985) for a zero value of the Reynolds number. The final example is a study of creeping flow of an incompressible viscous fluid in bearing geometries and is compared to work of Ingham and Kelmanson (1984).

The governing equation for steady, two-dimensional viscous flow of an incompressible fluid is written in terms of the stream function ψ and the vorticity ω as:

$$\nabla^4 \psi = R\left(\frac{\partial \psi}{\partial y} \frac{\partial \omega}{\partial x} - \frac{\partial \psi}{\partial x} \frac{\partial \omega}{\partial y} \right) \tag{5.13}$$

where R is the Reynolds number of the motion (Mills, 1977). This equation may be thought of as a nonhomogeneous biharmonic equation wherein the nonhomogeneous function (the right-hand side of Equation (5.13) is itself a nonlinear function of the field variables). Equation (5.13) may be transformed to an equivalent set

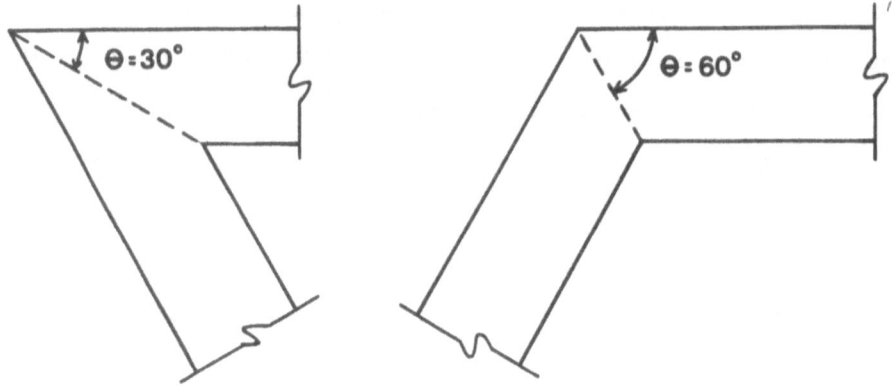

Figure 5.16 Corner Plates of Different Angles.

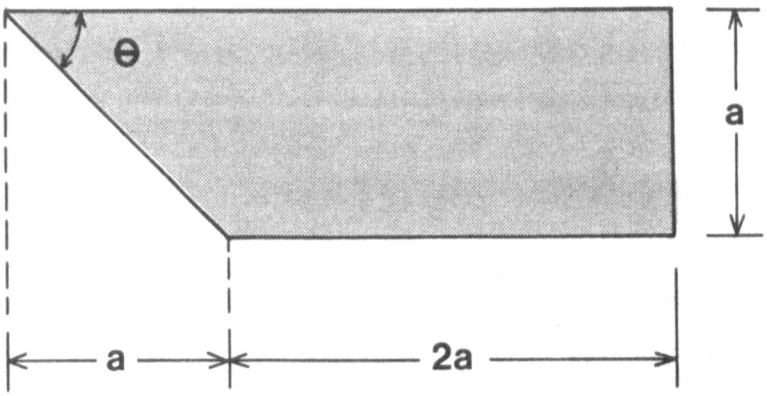

Figure 5.17 Corner Plate Problem Domain
Incorporating Symmetry.

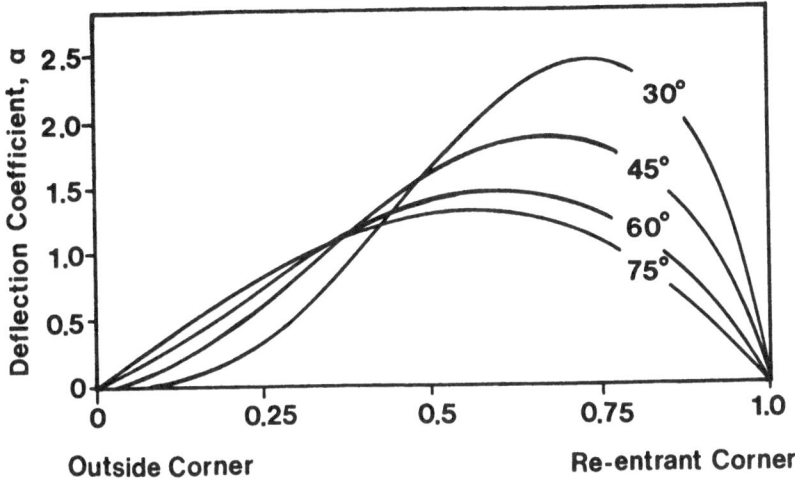

Figure 5.18   Deflection Across the Diagonal for Corner
              Plates of Different Angles.   Deflection
              $w = \alpha q b^4 (10^{-2})/D$.

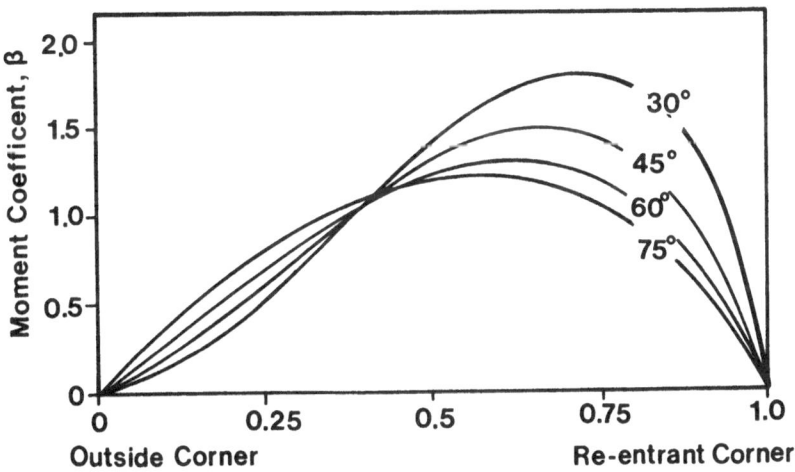

Figure 5.19   Moment Function Across the Diagonal for
              Corner Plates of Different Angles.
              Moment Function $M = \beta q b^2 (10^{-1})/D$.

of coupled Poisson type equations by introducing the relationship between the stream function and the vorticity. For non-zero values of the Reynolds number Equation (5.13) is solved using the iterative solution technique described in Chapter 4.

5.3.2 Inflow-Outflow Problem

The inflow-outflow problem considered in this example is defined as shown in Figure 5.20(a). The motion is generated by a viscous fluid entering and leaving the cylinder normal to the walls. The Reynolds number is defined in terms of the radius r, the entrance velocity U, the angle ε, and the kinematic viscosity ν, and given as $R = Ur\varepsilon/\nu$. In Figure 5.20(b), the solution of the flow field for a Reynolds number equal to zero is presented. The accuracy of this result, when checked by computing the exact infinite series solution given by Mills (1977), is excellent. As the Reynolds number increases, regions of recirculation develop as the flow becomes more unsteady. At the entrance and exit, the rate of change of the vorticity becomes large and the iterative solution technique will not converge to a physically correct solution.

5.3.3 Moving-Wall Problem

Shown in Figure 5.21(a) are the geometry and boundary conditions for a circular moving-wall problem. The fluid motion is generated by the rotation of part or all the boundary of the cylinder. This type of problem is important in the study of recirculating motion in cavities. The radius of the cylinder, r, the constant speed of the moving surface, U, and the kinematic viscosity, ν, will be used to define the Reynolds number as $R_1 = Ur/\nu$. Plots of the streamlines generated by the rotation of the upper half of the cylinder are shown in Figures 5.21(b), 5.22(a), and 5.22(b) for various Reynolds numbers. The flow at $R_1 = 0$ calculated from a closed form solution given in Mills (1977) and the numerical solution for the same flow conditions, shown in Figure 5.21(b), show excellent

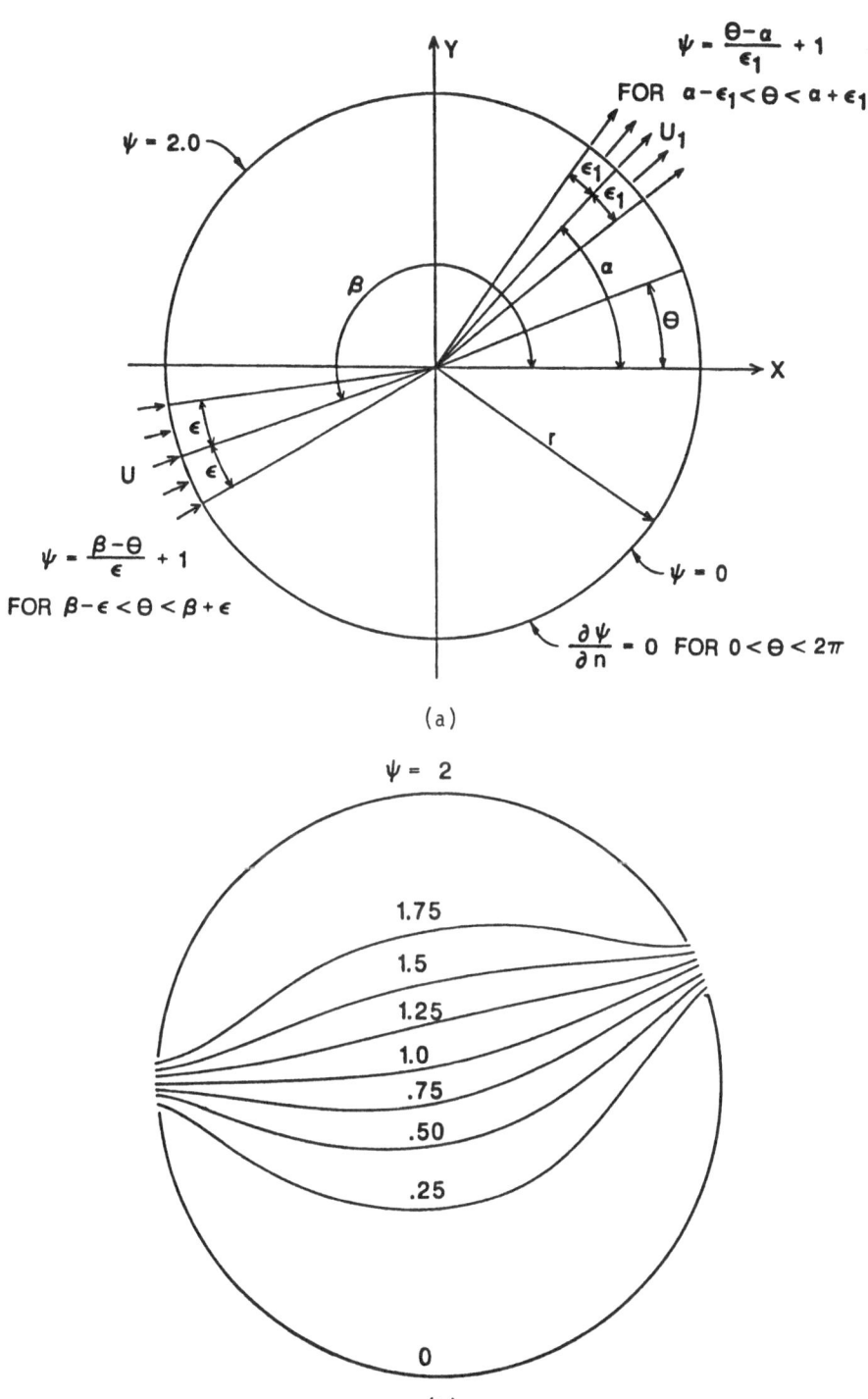

(a)

(b)

Figure 5.20 (a) Inflow-Outflow Problem Definition, r=2.0, α=π/8,
β=π, and ε=ε₁=π/32. (b) Streamlines for R=0.0.

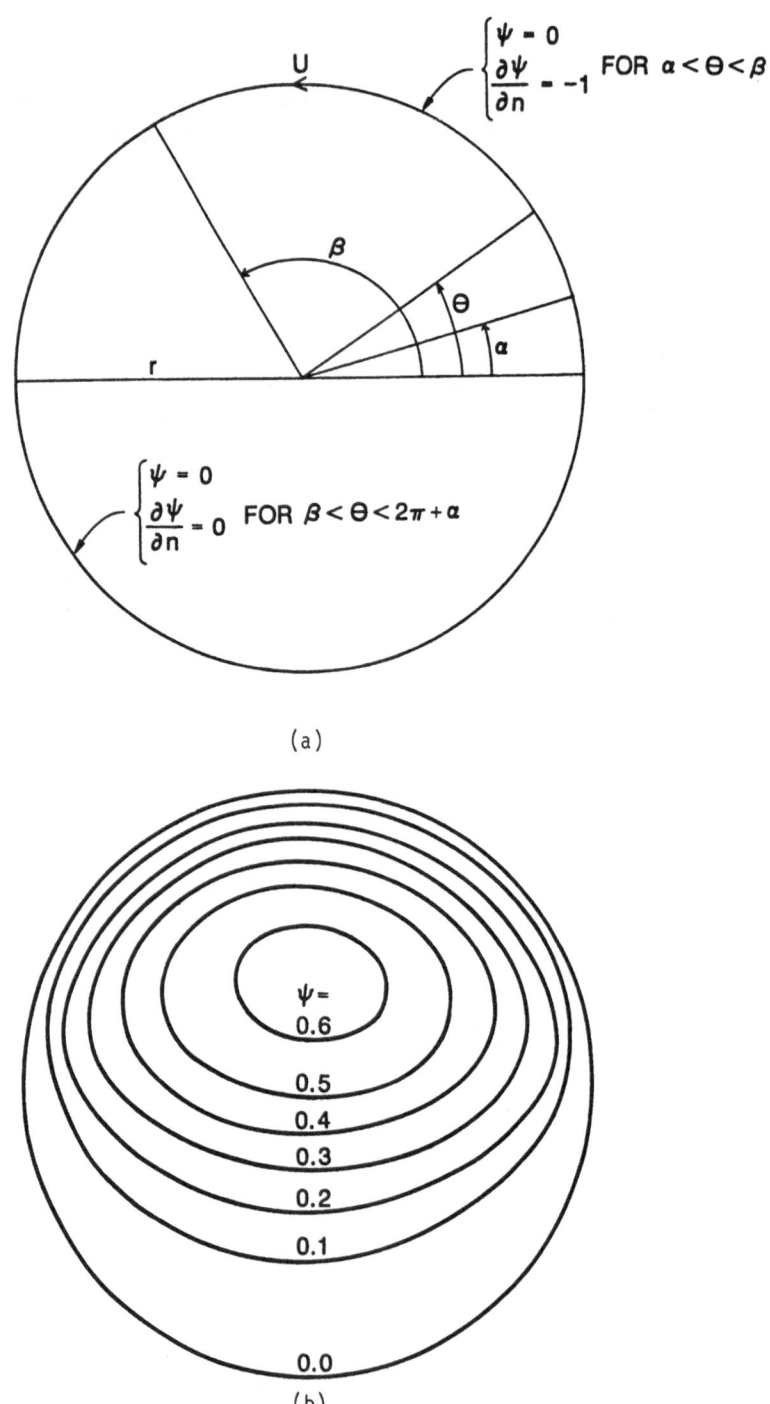

(a)

(b)

Figure 5.21   (a) Moving-Wall Problem Definition, r=2.0, α=0.0,
and β=π.   (b) Streamlines for R=0.0

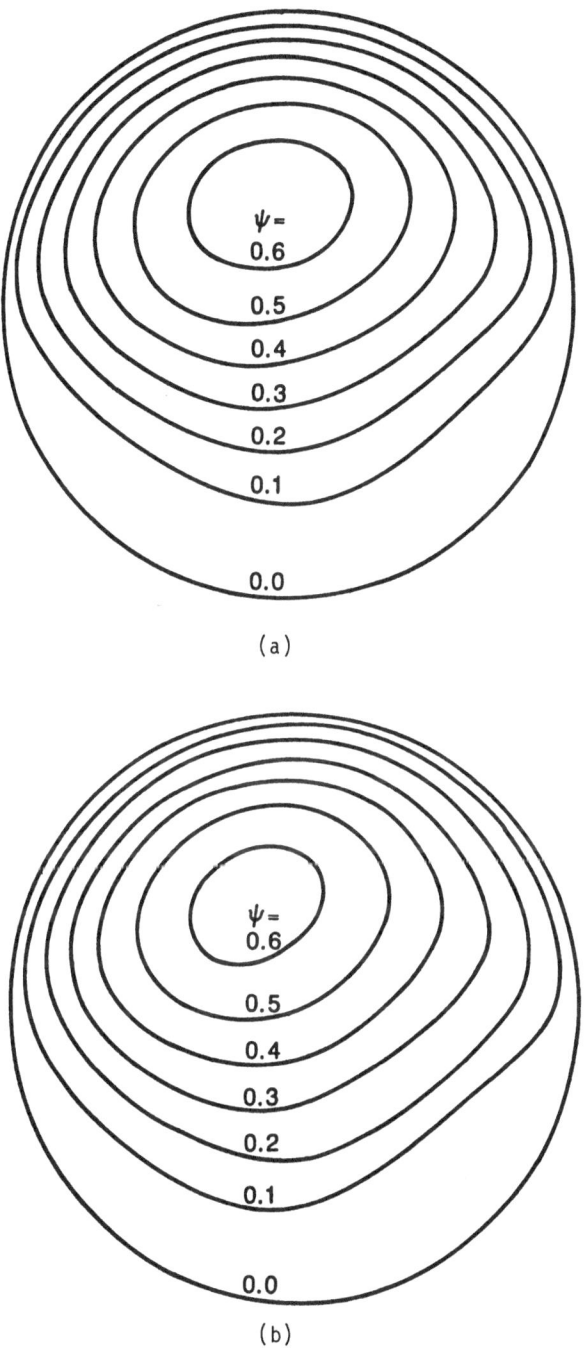

(a)

(b)

Figure 5.22 Streamline Plots for the Moving-Wall Problem,
(a) R_1=10.0, (b) R_1=20.0

agreement. Streamline plots for other values of Reynolds number compared favor-
ably with similar solutions presented by Mills (1977).

5.3.4 Flow Through a Fibrous Filter

Flow through an infinite array of cylinders is considered in this example.
A rectangular and a diagonal model of possible fiber configurations are presented. Symmetry reduces both problem geometries and boundary conditions to those
shown in Figures 5.23(a) and 5.23(b). In each case, the solution for a Reynolds
number of zero shows good agreement when compared to the results presented by
Hildyard et al (1985). Shown in Figures 5.24, 5.25, and 5.26 are plots of the
streamlines and vorticity contours for flows in the rectangular model characterized by Reynolds numbers of 0.0, 10.0 and 20.0 respectively. In Figures 5.27 and
5.28 streamline plots and vorticity contours for a Reynolds number of 0.0 and 2.0
are presented for flows in the diagonal model.

5.3.5 Flow in Bearings of Arbitrary Geometries

In this example, a slow incompressible viscous flow field in bearing geometries at zero Reynolds number is presented. The problem is defined by the region between an inner cylinder rotating at a constant angular velocity and an ou-
ter surface of arbitrary shape. The value of the stream function at the inner
cylinder, ψ_1, is an unknown constant. An additional equation for ψ_1 may be
obtained from the periodic nature of the pressure around the inner cylinder C_1
(Ingham and Kelmanson, 1984)

$$\int_{C_1} \frac{\partial P}{\partial S} \, dS = 0 \qquad (5.14)$$

Equation (5.14) may be rewritten in terms of the vorticity as

$$\int_{C_1} \frac{\partial \omega}{\partial n} \, dS = \int_{C_1} \omega' \, dS = 0 \qquad (5.15)$$

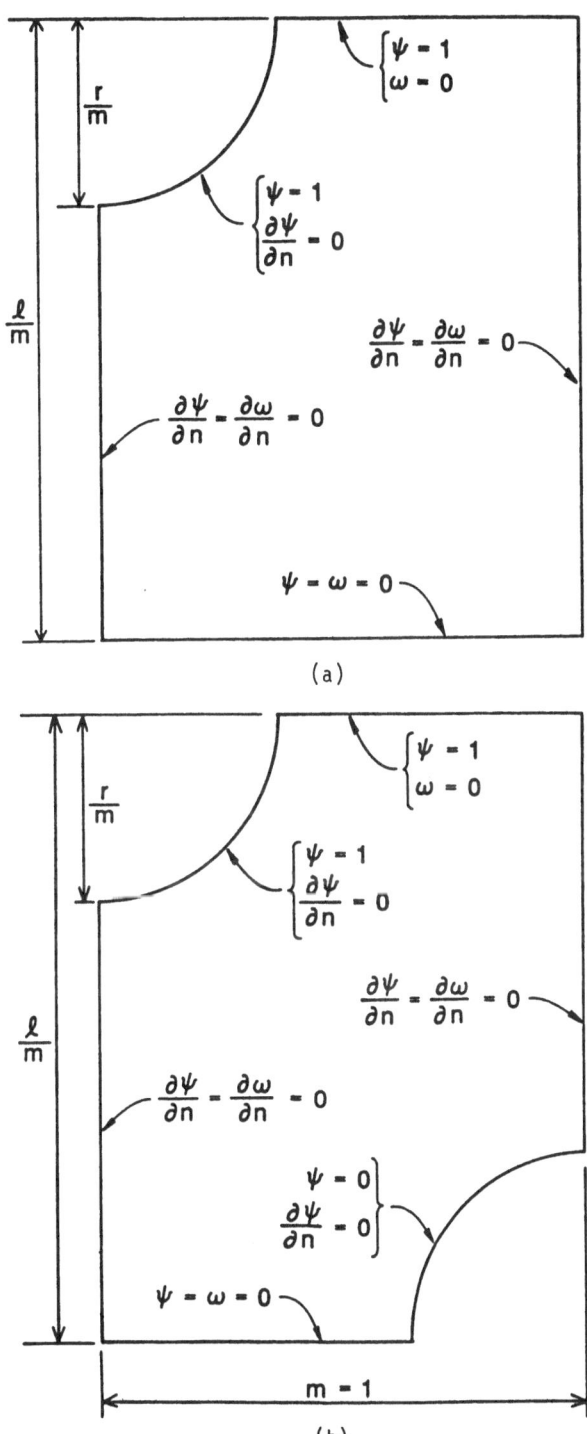

(a)

(b)

Figure 5.23 Problem Definition for Flow Through an Infinite Array of
Cylinders. (a) Rectangular Model; (b) Diagonal Model.

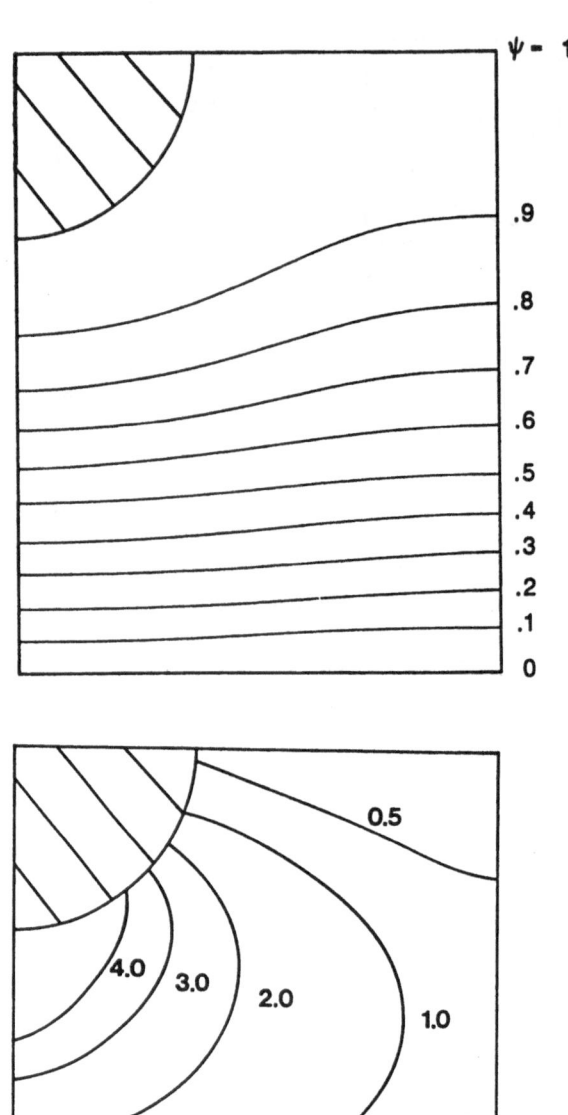

Figure 5.24   Streamline Plot and Vorticity Contours for Flow
Through an Infinite Reactangular Array of
Cylinders, R=0.0.

153

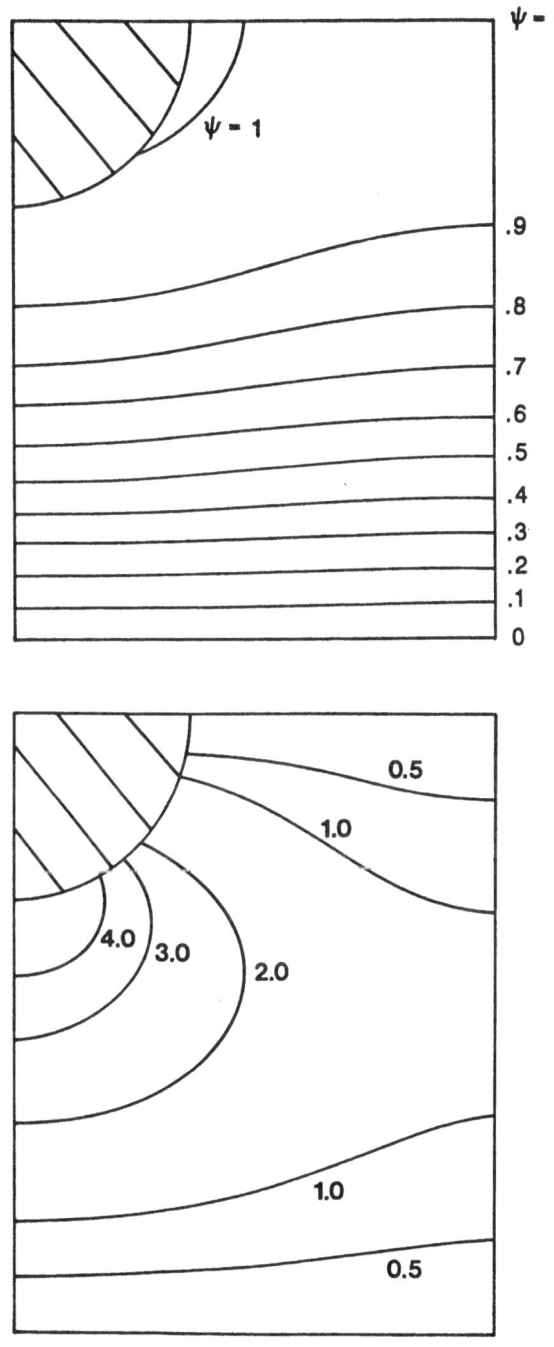

Figure 5.25 Streamline Plot and Vorticity Contours for Flow
Through an Infinite Rectangular Array of
Cylinders, R=10.0.

154

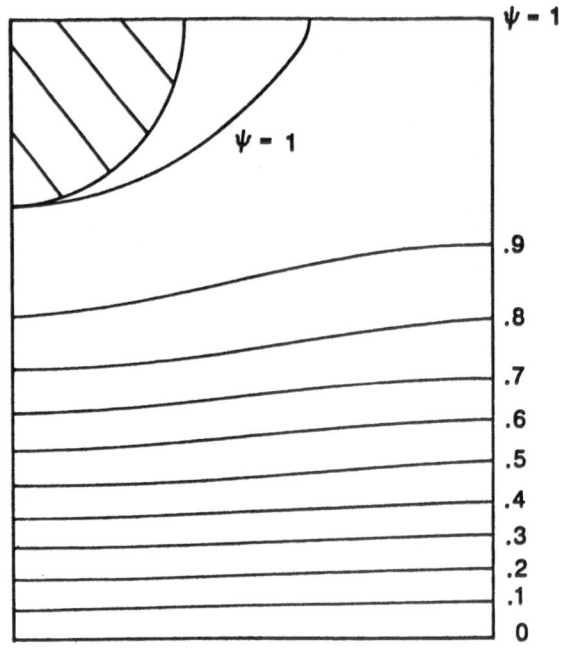

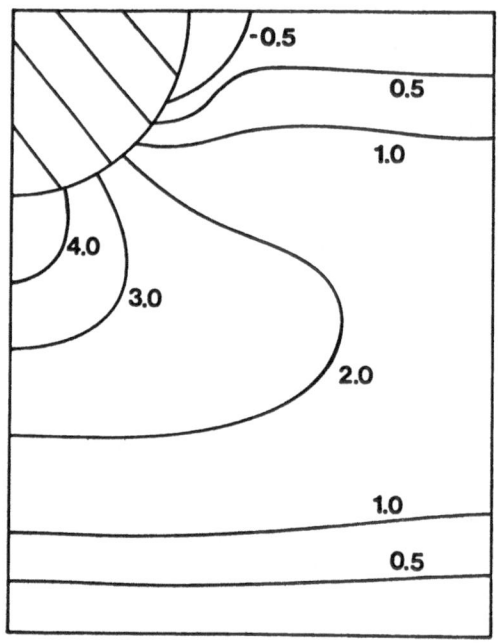

Figure 5.26   Streamline Plot and Vorticity Contours for Flow
Through an Infinite Rectangular   Array of
Cylinders, R=20.0.

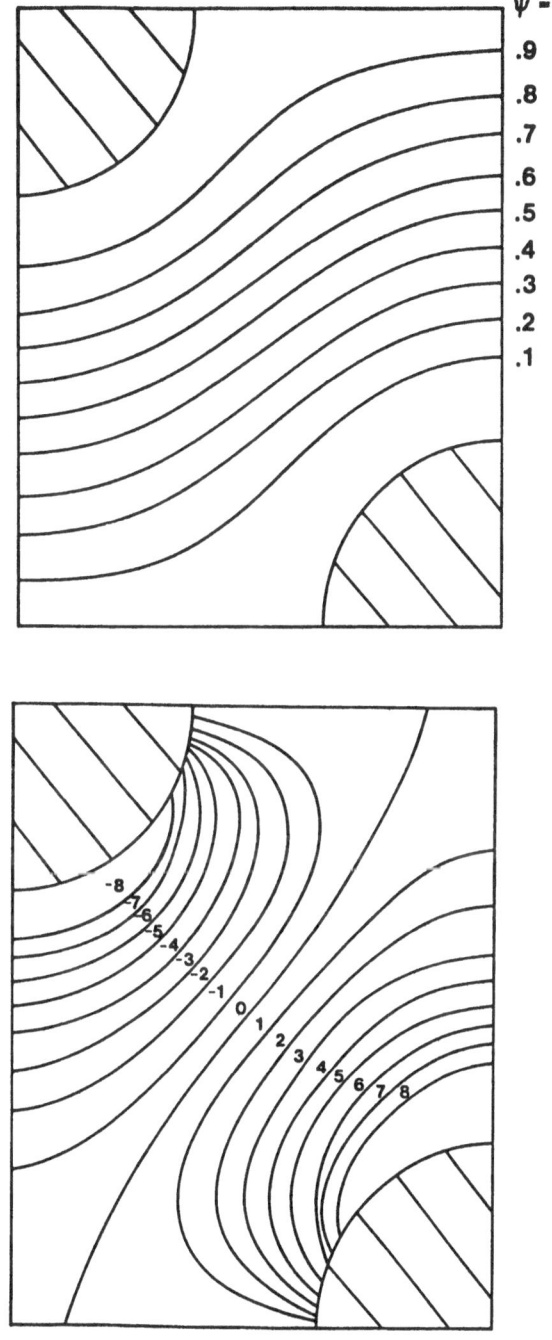

Figure 5.27 Streamline Plot and Vorticity Contours for Flow
Through an Infinite Diagonal Array of
Cylinders, R=0.0.

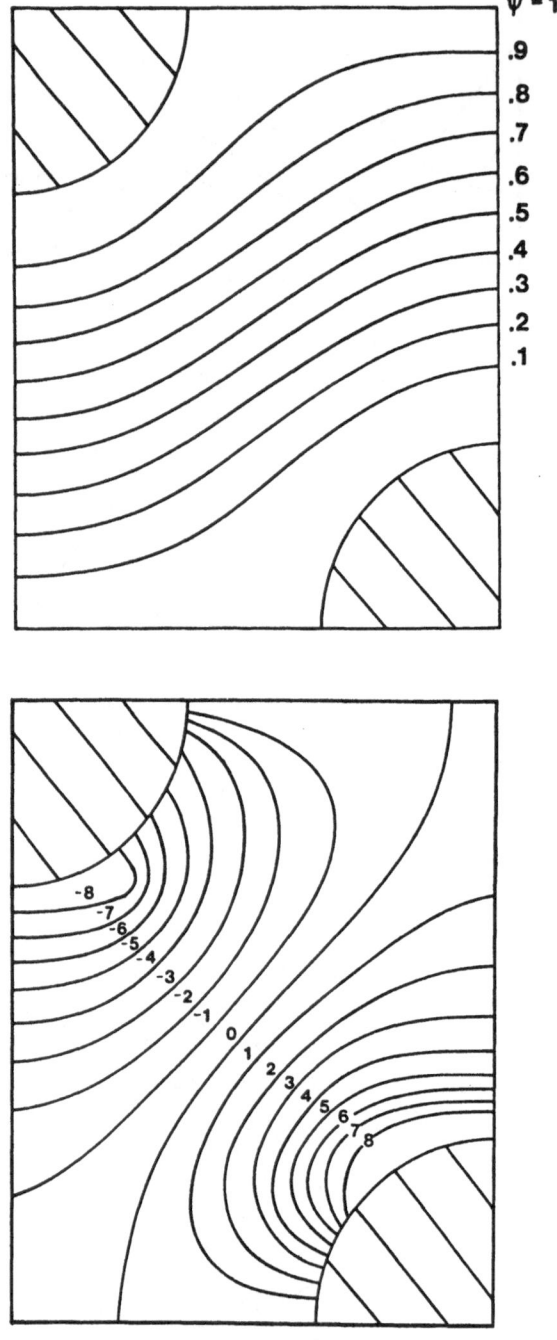

Figure 5.28 Streamline Plot and Vorticity Contours for Flow Through
an Infinite Diagonal Array of Cylinders, R=2.0.

Equation (5.15) provides the additional relationship required for a solution while accurately enforcing the pressure condition. Results given in the form of plots of streamlines and vorticity contours for cylindrical bearings with eccentricities of 0.5 and 0.8 are shown in Figures 5.29(a) and 5.29(b). Streamlines and vorticity contours for elliptical bearings with eccentricities of 0.5 and 0.8 are shown in Figures 5.30(a) and 5.30(b) respectively. Eccentricity for cylindrical geometries is defined as $e = \epsilon(r_2-r_1)$ and as $e = \epsilon(a_2-r_1)$ for elliptical bearings. In each case the results are in excellent agreement with those given by Ingham and Kelmanson (1984).

5.4 Concluding Remark

The examples presented in this chapter have demonstrated consistently that the boundary element formulation developed in this work accurately predicts the solution for a wide range of engineering problems of various geometries. In the next chapter, a description of the computer program used to work the non-iterative problems will be presented, as well as a complete FORTRAN listing of the program.

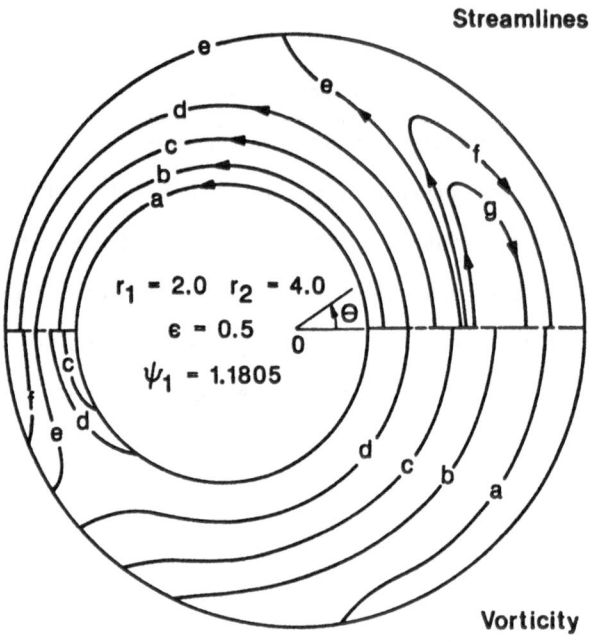

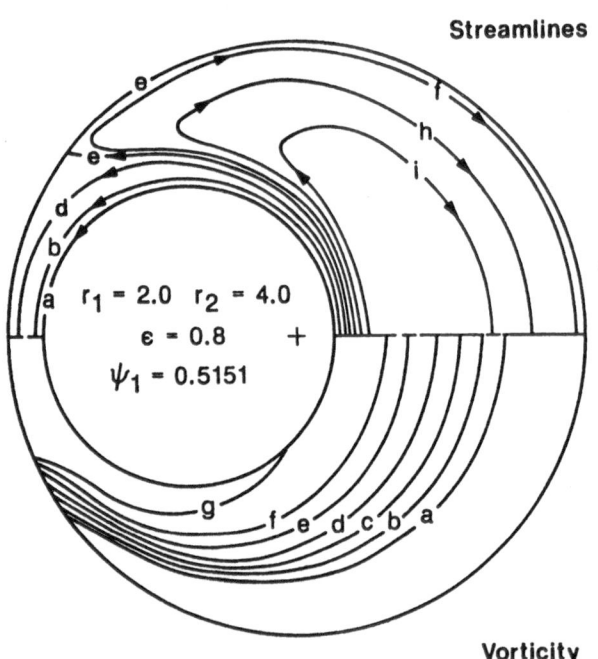

Figure 5.29 Streamlines and Vorticity Contours for an Eccentric
Bearing. Streamlines are at Values of ψ_1/N, where
N is (a) 1; (b) 1.5; (c) 3; (d) 10; (e) ∞; (f) -60;
(g) -30; (h) -5; (i) -2. Vorticities are at Values
of N Equal to (a) 0; (b) 0.3; (c) 0.6; (d) 1;
(e) 1.5; (f) 2 (g) 4.

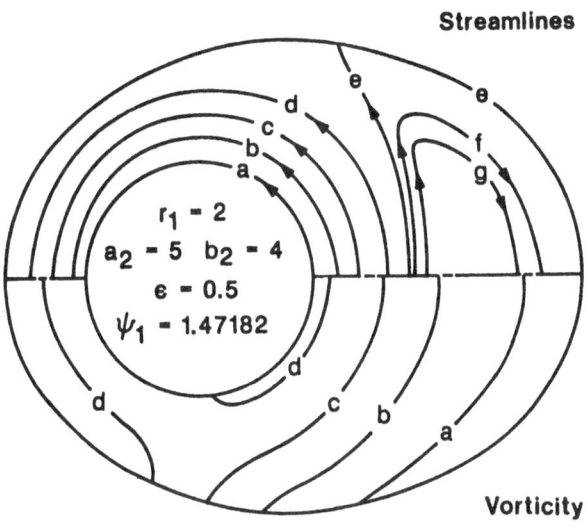

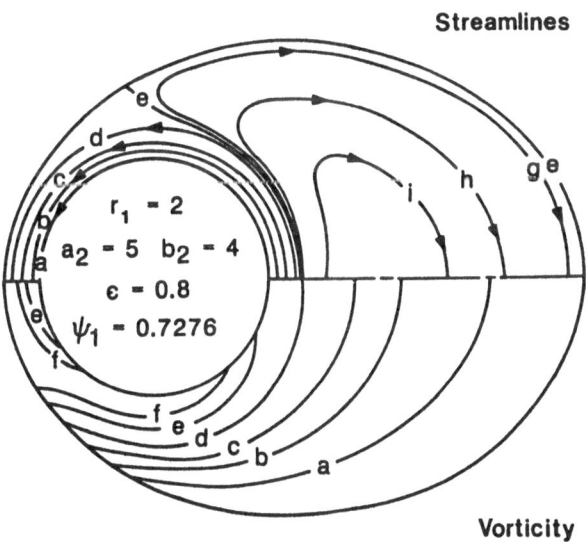

Figure 5.30   Streamlines and Vorticity Contours for an Elliptical
              Eccentric Bearing.   Streamlines are at Values of ψ_1/N,
              where N is (a) 1; (b) 1.5; (c) 3; (d) 10; (e) ∞;
              (f) -60; (g) -30 (h) -5; (i) -2.   Vorticities are
              at Values of N Equal to (a) 0; (b) 0.3; (c) 0.6;
              (d) 1; (e) 1.5; (f) 2.

CHAPTER 6

COMPUTER PROGRAM

6.1 Introduction

Based upon the theory presented in Chapters 2, 3, and 4 of this text, an improved boundary element formulation for the solution of the nonhomogeneous bi-harmonic equation was developed. The resulting computer code was designed in a modular style similar to that used by Gipson (1987). The structure should allow future modification without causing extensive rewriting of the code. The code is written in VAX FORTRAN 77 and should compile and run without any adjustments on any Digitial Equipment or VAX compatible machines. However, the code was written to be as machine independent as possible except for the interactive definitions of the input and output data sets in the main program. Conversion of the code to another brand of FORTRAN should be easily accomplished.

In this chapter, a computer listing of the formulation developed in this text will be presented in a segmental form. Each of the subroutines will be described separately with the intent of informing the user as to what each module does and how it works. In general, each subroutine is liberally commented on in order to point out important variables and calling arguments while providing insights on the execution of the formulation. In addition, a detailed user's guide and an example input data set are provided to clearly define the input data structure.

6.2 Main Program

The purpose of the main program is to define the structure and control the execution of each of the subroutines presented in this chapter. Although the main program performs a limited number of computations it serves to define the input/ output devices and the maximum dimensions of the problem (200 boundary nodes and 110 internal point locations in this version). The dimensions of the program may be adjusted to meet a particular user's needs and/or computer facilities by updating the appropriate numbers in the PARAMETER statement. A series of messages will automatically be printed to the standard output device to inform the user of the current stage of execution in the program.

```
C    =============================================================
C    =                                                           =
C    =                      P R O G R A M                        =
C    =                                                           =
C    =    PURPOSE --- TO DETERMINE THE SOLUTION OF THE NONHOMOGENEOUS  =
C    =                 BIHARMONIC EQUATION FOR TWO DIMENSIONAL   =
C    =                 GEOMETRIES                                =
C    =                                                           =
C    =    METHOD OF ANALYSIS --- THE BOUNDARY INTEGRAL EQUATION  =
C    =                   TECHNIQUE COUPLED WITH AN IMPROVED       =
C    =                   DOMAIN INTEGRATION SCHEME               =
C    =                                                           =
C    =                                                           =
C    =    ==> THIS PROGRAM IS WRITTEN IN VAX FORTRAN 77 AND SHOULD =
C    =        RUN WITHOUT MODIFICATIONS ON ANY DEC VAX PRODUCT   =
C    =        WITH APPROXIMATELY 2.5 MB OF MEMORY                =
C    =                                                           =
C    =============================================================
C
      IMPLICIT REAL*8(A-H,O-Z)
      PARAMETER (ND2=400,ND1=200,ND=110,NG=20)
      DIMENSION GKERN(ND2,ND2),HKERN(ND2,ND2),X(ND1),Y(ND1),
     $    ANSW(ND2),XCENT(ND),YCENT(ND),B(ND2),SOLV(ND2),
     $    FIXBND(ND2),NELE(ND1,4),KODE(ND1)
C
      COMMON/IO/INDEV,IOUTDV
      COMMON/MODEL/NG1,NTOT,NTOT2,NUMNP,NUMELE,NUMEQ,NUMINT,ITYPE,IPLOT
      COMMON/SOURCE/NSOUR,SLX(10),SLY(10),SVALUE(10)
C
C    =============================================================
C
      CHARACTER*10 NAME
      DATA INDEV,IOUTDV/15,16/
C
      WRITE(7,1)
1     FORMAT(//'$ENTER INPUT FILENAME :  ')
      READ(7,2) NAME
2     FORMAT(A10)
      OPEN(UNIT=INDEV,FILE=NAME,STATUS='OLD')
C
      WRITE(7,3)
3     FORMAT(//'$ENTER OUTPUT FILENAME :  ')
      READ(7,4) NAME
4     FORMAT(A10)
      OPEN(UNIT=IOUTDV,FILE=NAME,STATUS='NEW')
C
      NTOT=ND1
      NTOT2=ND2
      NG1=NG
```

```
C
C    ============================
C    =  INPUT OF PROGRAM DATA   =
C    ============================
C
      WRITE(7,31)
C
      CALL INPUT(XCENT,YCENT,X,Y,FIXBND,SOLV,KODE,NELE,NRANK)
C
C    ========================================
C    =     FORMATION OF SYSTEM EQUATIONS     =
C    ========================================
C
      WRITE(7,32)
C
      CALL FORMGH(X,Y,NELE,KODE,GKERN,HKERN,FIXBND,SOLV,B)
C
C    ========================================
C    =      SOLVE SYSTEM OF EQUATIONS        =
C    ========================================
C
      WRITE(7,33)
C
      CALL REDSOL(GKERN,NTOT2,ANSW,SOLV,NUMEQ,NRANK)
C
      DO 200 I=1,NUMEQ
  200   SOLV(I)=ANSW(I)
C
C    ==================================
C    =    OUTPUT OF COMPUTATIONS       =
C    ==================================
C
      CALL OUTPUT(X,Y,SOLV,FIXBND,XCENT,YCENT,NELE,KODE)
C
      WRITE(7,34)
C
   31 FORMAT(//1X,' DATA INPUT')
   32 FORMAT(1X,' FORMING COEFFICIENTS USING FORMGH')
   33 FORMAT(1X,' SOLVING USING REDSOL')
   34 FORMAT(1X,' OUTPUT OF RESULTS',//)
C
      CALL EXIT
      END
```

6.3 Subroutine INPUT

INPUT reads the input data describing the problem in question into the program memory and echoes this information to the defined output device. The initial program input parameters control the type of biharmonic analysis to be performed, the number of discrete boundary nodes and elements desired to define the problem geometry, the number of internal point locations where information on the field function and its derivatives are required, and the number, location, and strength of any concentrated source points.

The user may choose between a homogeneous or nonhomogeneous biharmonic analysis. Several different methods to evaluate the nonhomogeneous source term $f(x,y)$, are available. The first is an exact analysis for a specialized form of $f(x,y)$ defined in Equation (4.21). In this version of the program, the method is available for constant and linear elements only. If this form of analysis is chosen, some additional input data is required defining the form of the function $f(x,y)$. The second is a technique capable of evaluating $f(x,y)$, for all element types, if it is itself a biharmonic function. The last type of analysis can handle an arbitrary form of $f(x,y)$ using a domain "fanning" algorithm.

After the initial problem parameters are input, the boundary nodes are identified by a number, their coordinates, and a boundary condition code. To simplify node data input, only the end nodes of a rectilinear segment need be defined. Omitted node numbers and coordinates are uniformly generated between the two end nodes in a straight line. The generated nodal points have the same boundary condition information as the first end node defining the segment. This version of the program allows up to 200 nodes.

Each internal point location where the value of the field function and/or its derivatives is desired is sequentially numbered. Omitted internal point numbers have their coordinates generated in a manner similar to the boundary node data. The program is currently dimensioned to allow up to 110 internal points.

The elements describing the problem geometry are numbered in ascending order counterclockwise around the domain. The user may specify any one of the four element types in a variety of combinations by simply indicating the node numbers making up that element. Similar to the boundary node input structure, omitted elements have their node numbers uniformly generated consistent with the first and last elements of the generation sequence.

```
      SUBROUTINE INPUT(XCENT,YCENT,X,Y,FIXBND,SOLV,KODE,NELE,NRANK)
C
C =================================================================
C =                                                               =
C =          S U B R O U T I N E      I N P U T                   =
C =                                                               =
C = THIS SUBROUTINE READS THE INPUT DATA                          =
C = CALLING ARGUMENTS -                                           =
C =                                                               =
C =  XCENT,YCENT - VECTORS CONTAINING COORDINATES OF INTERNAL     =
C =                POINTS WHERE THE USER DESIRES THE FUNCTION      =
C =                TO BE CALCULATED                                =
C =  FIXBND - VECTOR OF FIXED BOUNDARY CONDITIONS                  =
C =  KODE - VECTOR CONTAINING BOUNDARY CONDITION CODES             =
C =                                                               =
C =================================================================
C
      IMPLICIT REAL*8 (A-H,O-Z)
      DIMENSION TITLE(18),XCENT(1),YCENT(1),X(1),Y(1),FIXBND(1),
     $         SOLV(1),NELE(NTOT,4),KODE(1)
      COMMON/IO/INDEV,IOUTDV
      COMMON/MODEL/NG1,NTOT,NTOT2,NUMNP,NUMELE,NUMEQ,NUMINT,ITYPE,IPLOT
      COMMON/SOURCE/NSOUR,SLX(10),SLY(10),SVALUE(10)
      COMMON/PCON/C1,C2,C3,C4
      COMMON/IHOEL/IHIGH
      DATA SMALL /1.0E-10/
C
C =================================================================
C =                                                               =
C = TITLE AND CONTROL CARDS                                       =
C = TITLE(18A4) - HEADING CARD CONTAINING USER'S                  =
C =               DESCRIPTIVE INFORMATION                         =
C = CONTROL CARD(4I5) - CONTAINS THE FOLLOWING PARAMETERS         =
C =      ITYPE = 0 LINEAR ANALYSIS                                =
C =            = 1 EXACT HARMONIC SOURCE TERM (LINEAR OR          =
C =                 CONSTANT ELEMENTS ONLY)                       =
C =            = 2 BIHARMONIC SOURCE TERM                         =
C =            = 3 ARBITRARY SOURCE TERM                          =
C =                                                               =
C =      NUMNP  = TOTAL NUMBER OF NODAL POINTS( 200 MAXIMUM )     =
C =      NUMELE = TOTAL NUMBER OF ELEMENTS                        =
C =      NSOUR  = TOTAL NUMBER OF POINT SOURCES                   =
C =                                                               =
C =   ==> IF ITYPE = 0, THEN THE VALUE OF THE COEFFICIENTS OF THE =
C =                  SPECIAL HARMONIC SOURCE TERM f(x,y) ARE      =
C =                  INPUT: f(x,y)=C1xy + C2x + C3y + C4          =
C =                                                               =
C =   ==> IF NSOUR > 0, THEN THE LOCATION (SLX(I),SLY(I))         =
C =                  AND THE STRENGTH SVALUE(I) OF EACH           =
C =                  POINT SOURCE IS SPECIFIED                    =
C =                                                               =
C =================================================================
C
      READ(INDEV,1000)TITLE
      READ(INDEV,1100)ITYPE,NUMNP,NUMELE,NUMINT,NSOUR
      IF(NSOUR.GE.1)READ(INDEV,1105)(SLX(I),SLY(I),SVALUE(I),I=1,NSOUR)
      IF(ITYPE.EQ.1)READ(INDEV,1110)C1,C2,C3,C4
C
      NUMEQ=NUMNP+NUMNP
C
      WRITE(IOUTDV,2000)TITLE,ITYPE,NUMNP,NUMELE,NUMINT
      IF(ITYPE.EQ.0)WRITE(IOUTDV,2110)
      IF(ITYPE.EQ.1)WRITE(IOUTDV,2120)C1,C2,C3,C4
      IF(ITYPE.EQ.2)WRITE(IOUTDV,2130)
      IF(ITYPE.EQ.3)WRITE(IOUTDV,2140)
      IF(NSOUR.GE.1)THEN
        WRITE(IOUTDV,2160)
        WRITE(IOUTDV,2165)(SVALUE(I),SLX(I),SLY(I),I=1,NSOUR)
      END IF
```

```
C
C   ================================================================
C   =                                                              =
C   =  INPUT OF BOUNDARY COORDINATES (EXTREME POINTS OF            =
C   =  BOUNDARY ELEMENTS) AND BOUNDARY CONDITIONS (2I5,4F15.0)     =
C   =  OMITTED NODES ARE GENERATED IN A STRAIGHT LINE              =
C   =                                                              =
C   =  - NUMBERING OF NODES PROCEEDS IN ASCENDING ORDER            =
C   =    NODE(I) = NODE NUMBER                                     =
C   =    KODE(I) = BOUNDARY CONDITION CODE                         =
C   =            = 1 IF NORMAL DERIVATIVE OF FUNCTION IS SPECIFIED  =
C   =                AT THE NODE                                   =
C   =            = 2 IF FUNCTION IS SPECIFIED AT NODE              =
C   =            = 3 IF NORMAL DERIVATIVE OF THE LAPLACIAN OF THE   =
C   =                FUNCTION IS SPECIFIED AT THE NODE             =
C   =            = 4 IF THE LAPLACIAN OF THE FUNCTION IS SPECIFIED  =
C   =                AT THE NODE                                   =
C   =    *** NOTE ***  KODE(I) IS A TWO DIGIT NUMBER SINCE TWO     =
C   =    BOUNDARY CONDITIONS MUST BE SPECIFIED AT EACH NODE        =
C   =    EXAMPLES: KODE(1)=13 MEANS AT NODE ONE BOTH KODE=1 AND    =
C   =              KODE=3 ARE SPECIFIED                            =
C   =              POSSIBLE KODE SPECIFICATIONS: 12,13,14,23,24    =
C   =                                                              =
C   =    X(I) = GLOBAL X-COORDINATE OF NODE                        =
C   =    Y(I) = GLOBAL Y-COORDINATE OF NODE                        =
C   =    FIXBND(I) = BOUNDARY CONDITIONS AT NODE                   =
C   =                                                              =
C   =    GENERATED BOUNDARY CONDITIONS ARE DISTRIBUTED LINEARLY    =
C   =    BETWEEN THE VALUES SPECIFIED ON THE TWO INPUT LINES IN    =
C   =    SEQUENCE                                                  =
C   =                                                              =
C   ================================================================
C
      IERROR=0
      WRITE(IOUTDV,2200)
      L=1
   60 READ(INDEV,1200)N,KODE(N),X(N),Y(N),FIXBND(N),FIXBND(N+NUMNP)
      DIFF=DFLOAT(N+1-L)
      IF(N-L)80,110,90
   80 WRITE(IOUTDV,2210)N
      IERROR=1
C
      GO TO 60
C
   90 DX=(X(N)-X(L-1))/DIFF
      DY=(Y(N)-Y(L-1))/DIFF
      DFIX1=(FIXBND(N)-FIXBND(L-1))/DIFF
      DFIX2=(FIXBND(N+NUMNP)-FIXBND(L-1+NUMNP))/DIFF
C
  100 KODE(L)=KODE(N)
      X(L)=X(L-1)+DX
      Y(L)=Y(L-1)+DY
      FIXBND(L)=FIXBND(L-1)+DFIX1
      FIXBND(L+NUMNP)=FIXBND(L-1+NUMNP)+DFIX2
C
  110 WRITE(IOUTDV,2220)L,KODE(L),X(L),Y(L)
      L=L+1
      IF(N-L)120,110,100
  120 IF(NUMNP+1.GT.L)GO TO 60
C
      WRITE(IOUTDV,2225)
      DO 10 I=1,NUMNP
      N1=I+NUMNP
      KN=KODE(I)
      SOLV(I)=0.0
      SOLV(N1)=0.0
C
      IF(KODE(I).EQ.12)
     $  WRITE(IOUTDV,2230)I,KN,FIXBND(I),FIXBND(N1),SOLV(I),SOLV(N1)
```

```
      IF(KODE(I).EQ.13)
     $   WRITE(IOUTDV,2230)I,KN,FIXBND(I),SOLV(I),FIXBND(N1),SOLV(N1)
      IF(KODE(I).EQ.14)
     $   WRITE(IOUTDV,2230)I,KN,FIXBND(I),SOLV(I),SOLV(N1),FIXBND(N1)
      IF(KODE(I).EQ.23)
     $   WRITE(IOUTDV,2230)I,KN,SOLV(I),FIXBND(I),FIXBND(N1),SOLV(N1)
      IF(KODE(I).EQ.24)
     $   WRITE(IOUTDV,2230)I,KN,SOLV(I),FIXBND(I),SOLV(N1),FIXBND(N1)
C
  10   CONTINUE
C
      IF(NUMINT.EQ.O)GOTO 300
C
      WRITE(IOUTDV,2250)
      L=1
  160 READ(INDEV,1300)N,XCENT(N),YCENT(N)
      DIFF=DFLOAT(N+1-L)
      IF(N-L)180,210,190
  180 WRITE(IOUTDV,2210)N
      IERROR=1
      GO TO 160
  190 DX=(XCENT(N)-XCENT(L-1))/DIFF
      DY=(YCENT(N)-YCENT(L-1))/DIFF
  200 XCENT(L)=XCENT(L-1)+DX
      YCENT(L)=YCENT(L-1)+DY
  210 WRITE(IOUTDV,2300)L,XCENT(L),YCENT(L)
      L=L+1
      IF(N-L)220,210,200
  220 IF(NUMINT+1.GT.L)GO TO 160
C
C  ===================================================================
C  =                                                                 =
C  =   INPUT OF ELEMENT CARDS(4I5)                                    =
C  =                                                                 =
C  = - ELEMENTS ARE NUMBERED IN ASCENDING ORDER COUNTERCLOCKWISE      =
C  =     AROUND INTERIOR OF REGION(CLOCKWISE FOR AN EXTERIOR REGION)  =
C  =                              .                                  =
C  ===> FOR CONSTANT ELEMENTS: (N1,N2,N3)                             =
C  =       N1,N2 = COORDINATES DEFINING THE ELEMENT;                  =
C  =               N2 IS LOCATED IN COUNTERCLOCKWISE POSITION         =
C  =                  WITH RESPECT TO N1                              =
C  =          N3 = MUST BE A NEGATIVE NUMBER TO INDICATE A            =
C  =               CONSTANT ELEMENT                                   =
C  =                                                                 =
C  ===> FOR LINEAR ELEMENTS: (N1,N2)                                  =
C  =       N1,N2 = NODE NUMBERS DEFINING THE ELEMENT                  =
C  =                                                                 =
C  ===> FOR QUADRATIC ELEMENTS: (N1,N2,N3)                            =
C  =       N1,N2,N3 = NODE NUMBERS DEFINING THE END NODE, N1, THE     =
C  =                  MIDDLE NODE, N2, AND THE SECOND END NODE, N3,   =
C  =                  LOCATED IN A COUNTERCLOCKWISE POSITION FROM N1  =
C  =                                                                 =
C  ===> FOR OVERHAUSER ELEMENTS: (N1,N2,N3,N4)                        =
C  =       N1,N2,N3,N4 = NODE NUMBERS DEFINING THE FIRST "EXTRA" =
C  =                     NODE N1, THE FIRST "REGULAR" NODE, N2, AND=
C  =                     THE SECOND COUNTERCLOCKWISE OCCURANCE       =
C  =                     OF THE NEXT "REGULAR" NODE, N3, AND THE     =
C  =                     NEXT "EXTRA" NODE, N4                        =
C  =                                                                 =
C  =     OMITTED ELEMENT CARDS ARE GENERATED; NODES ARE GENERATED     =
C  =     IN EVENLY SPACED INCREMENTS CONSISTENT WITH THE FIRST        =
C  =     AND LAST CARDS IN THE GENERATION SEQUENCE                    =
C  =                                                                 =
C  ===================================================================
C
      IHIGH=O
  300 WRITE(IOUTDV,2450)
      L=1
C
```

```
 360   READ(INDEV,1450)IEL,N1,N2,N3,N4
       IF(IHGIH.EQ.O)THEN
         IF(N3.NE.O)IHIGH=1
         IF(N4.NE.O)IHIGH=2
       ENDIF
       NDIF=IEL+1-L
       IF(IEL-L)380,385,390
 380   WRITE(IOUTDV,2210)IEL
       IERROR=1
       GO TO 360
 390   ND1=(N1-N11)/NDIF
       ND2=(N2-N22)/NDIF
       ND3=(N3-N33)/NDIF
       ND4=(N4-N44)/NDIF
 400   N1=N11+ND1
       N2=N22+ND2
       N3=N33+ND3
       N4=N44+ND4
 385   WRITE(IOUTDV,2610)L,N1,N2,N3,N4
       NELE(L,1)=N1
       NELE(L,2)=N2
       NELE(L,3)=N3
       NELE(L,4)=N4
       N11=N1
       N22=N2
       N33=N3
       N44=N4
       L=L+1
C
       IF(IEL-L)410,400,400
C
 410   IF(NUMELE+1.GT.L)GO TO 360
C
C  ==============================
C  =  DETERMINE RANK OF MATRIX  =
C  ==============================
C
       NSUM=O
       DO 500 I=1,NUMNP
         XP=X(I)
         YP=Y(I)
       DO 500 J=1,NUMNP
         IF(I.EQ.J)GOTO 500
           XG=X(J)
           YG=Y(J)
C
         IF(DABS(XP-XG).LT.SMALL.AND.DABS(YP-YG).LT.SMALL)THEN
C
           IF(KODE(I).EQ.12)THEN
             IF(KODE(J).EQ.13)NSUM=NSUM+1
             IF(KODE(J).EQ.14)NSUM=NSUM+1
             IF(KODE(J).EQ.23)NSUM=NSUM+1
             GOTO 500
           END IF
C
           IF(KODE(I).EQ.13)THEN
             IF(KODE(J).EQ.12)NSUM=NSUM+1
             IF(KODE(J).EQ.14)NSUM=NSUM+1
             IF(KODE(J).EQ.23)NSUM=NSUM+1
             GOTO 500
           END IF
C
           IF(KODE(I).EQ.14)THEN
             IF(KODE(J).EQ.12)NSUM=NSUM+1
             IF(KODE(J).EQ.13)NSUM=NSUM+1
             IF(KODE(J).EQ.24)NSUM=NSUM+1
             GOTO 500
           END IF
C
```

```
          IF(KODE(I).EQ.23)THEN
            IF(KODE(J).EQ.12)NSUM=NSUM+1
            IF(KODE(J).EQ.13)NSUM=NSUM+1
            IF(KODE(J).EQ.24)NSUM=NSUM+1
            GOTO 500
          END IF
C
          IF(KODE(I).EQ.24)THEN
            IF(KODE(J).EQ.23)NSUM=NSUM+1
            IF(KODE(J).EQ.14)NSUM=NSUM+1
            GOTO 500
          END IF
        END IF
C
 500    CONTINUE
C
        NRANK=NUMEQ-(NSUM/2)
C
        RETURN
C
C  =======================
C  =  FORMAT STATEMENTS  =
C  =======================
C
 1000 FORMAT (20A4)
 1100 FORMAT (5I5)
 1105 FORMAT (3F10.3)
 1110 FORMAT (4F10.2)
 1200 FORMAT (2I5,4F15.0)
 1300 FORMAT (I10,2F15.0)
 1450 FORMAT (5I5)
 2000 FORMAT (80('*')//1X,18A4//80('*')//
     1        10X,'PROBLEM TYPE---------------',I4/
     2        10X,'NUMBER OF NODAL POINTS-----',I4/
     2        10X,'NUMBER OF ELEMENTS---------',I4/
     3        10X,'NUMBER OF INTERNAL POINTS--',I4/)
 2110 FORMAT (1X,' HOMOGENEOUS BIHARMONIC EQUATION'/)
 2120 FORMAT (//1X,' LINEAR BIHARMONIC EQUATION; HARMONIC SOURCE TERM'//,
     1        1X,' F(X,Y) = C1 * XY + C2 * X + C3 * Y + C4',//
     2        11X,'C1=',F10.4,5X,'C2=',F10.4,/
     3        11X,'C3=',F10.4,5X,'C4=',F10.4//)
 2130 FORMAT (1X,' LINEAR BIHARMONIC EQUATION; BIHARMONIC SOURCE TERM'/,
     1        '  DEFINED IN FUNCTION STATEMENTS IN SUBROUTINE FUNCTION'//)
 2140 FORMAT (1X,' LINEAR BIHARMONIC EQUATION; GENERAL SOURCE TERM'/,
     1        '  DEFINE SOURCE TERM IN SUBROUTINE FUNCTION'//)
 2160 FORMAT (1X,' POINT SOURCE TERMS',/,1X,20('-'),/,
     1        1X,'    STRENGTH    LOCATION (X,Y)'/)
 2165 FORMAT (1X,F10.2,5X,'(',F7.2,',',F7.2,')')
 2200 FORMAT (//,1X,'NODE #',8X,'CODE',4X,'X-COORDINATE',4X,
     1        'Y-COORDINATE'/,80('-')/)
 2210 FORMAT (10HOCARD NO. I4, 13H OUT OF ORDER )
 2220 FORMAT (I5,5X,I8,2X,2E15.6)
 2225 FORMAT (//80('-')//,1X,'NODE # CODE',11X,'F',14X,'F¢',11X,
     1        'LP(F)',10X,'LP(F)¢'/)
 2230 FORMAT (1X,I4,I7,4F15.3)
 2250 FORMAT (//80('-')//10X,'INTERNAL POINTS -----'//,26X,'X',13X,
     1        'Y'/17X,35('-')/)
 2300 FORMAT (I10,5X,2F15.6)
 2450 FORMAT (/80('-')//7X,'ELEMENT NO.      NI     NJ     NK     NL',
     1        /6X,43('-')//)
 2610 FORMAT (10X,I3,11X,I3,3(4X,I3))
C
      END
```

6.4 Subroutine FORMGH

FORMGH is one of the most important routines in this program. Its first
task is the construction and collocation of the [G], [H], [K], and [L] matrices
described in Chapter 2. The subroutine has the inherent ability to distinguish
between element types and assemble each matrix accordingly. The routine sorts
the matrices consistent with the prescribed boundary conditions. For non-zero
forms of the source function f(x,y) the subroutine forms the vectors {B1} and
{B2}. Before exiting to the main program, FORMGH creates a system of equations
of the form [A] {Y} = {F}, ready for solution of the unknown boundary
quantities. Another important function of this routine is the generation of
Gaussian quadrature points and weights necessary for any subsequent numerical
integration. The number of quadrature points may be adjusted by the user by
updating the variable NUMGP in subroutine FORMGH.

```
      SUBROUTINE FORMGH(X,Y,NELE,KODE,GKERN,HKERN,FIXBND,SOLV,B)
C
C     ================================================================
C     =                                                              =
C     =  THIS SUBROUTINE FORMS THE SYSTEM MATRICES AND ARRANGES      =
C     =  THE EQUATIONS TO BE SOLVED                                  =
C     =                                                              =
C     =  CALLING ARGUMENTS -                                         =
C     =    X,Y - VECTORS CONTAINING THE NODAL COORDINATES            =
C     =    GKERN,HKERN - ARRAYS CONTAINING THE SYSTEM COEFFICIENTS   =
C     =                  TO BE CALCULATED IN THIS ROUTINE            =
C     =    FIXBND - VECTOR OF BOUNDARY CONDITIONS                    =
C     =    KODE - VECTOR OF BOUNDARY CONDITION CODES                 =
C     =    SOLV - VECTOR WHICH WILL CONTAIN SOLUTION SET             =
C     =    B - VECTOR TO CONTAIN POISSON TERMS                       =
C     =                                                              =
C     ================================================================
C
      IMPLICIT REAL*8 (A-H,O-Z)
      REAL*8 KI,KJ,KK,KL,LI,LJ,LK,LL
      COMMON/IO/INDEV,IOUTDV
      COMMON/MODEL/NG1,NTOT,NTOT2,NUMNP,NUMELE,NUMEQ,NUMINT,ITYPE,IPLOT
      COMMON/GQUAD/GP(100),GW(100),NUMGP
      COMMON/IHOEL/IHIGH
      COMMON/SOURCE/NSOUR,SLX(10),SLY(10),SVALUE(10)
      DIMENSION GKERN(NTOT2,NTOT2),HKERN(NTOT2,NTOT2),NELE(NTOT,4),
     $ SOLV(1),X(1),Y(1),FIXBND(1),KODE(1),B(1)
      DATA EPS/3.D-16/
      DATA PI/3.1415926535897932/
C
      DO 1 I=1,NUMEQ
        B(I)=0.0
    1 CONTINUE
```

```
C
      DO 2 I=1,NUMEQ
      DO 2 J=1,NUMEQ
        GKERN(I,J)=0.0
        HKERN(I,J)=0.0
 2    CONTINUE
C
      IF(ITYPE.GE.2.OR.IHIGH.GE.1)THEN
C
C ================================================
C *  GENERATE NUMGP NUMBER OF GAUSSIAN POINTS   =
C *     OVER AN INTERVAL OF O<=X<=1             =
C ================================================
C
      NUMGP=16
C
      M=(NUMGP+1)/2
      XM=0.5
      XL=0.5
C
        DO 70 I=1,M
          Z=COS(PI*(I-0.25)/(NUMGP+0.5))
 90       CONTINUE
            P1=1.0
            P2=0.0
              DO 80 J=1,NUMGP
                P3=P2
                P2=P1
                P1=((2.0*J-1.0)*Z*P2-(J-1.0)*P3)/J
 80           CONTINUE
            PP=NUMGP*(Z*P1-P2)/(Z*Z-1.0)
            Z1=Z
            Z=Z1-P1/PP
          IF(DABS(Z-Z1).GT.EPS)GOTO 90
          GP(I)=XM-XL*Z
          GP(NUMGP+1-I)=XM+XL*Z
          GW(I)=2.0*XL/((1.0-Z*Z)*PP*PP)
          GW(NUMGP+1-I)=GW(I)
 70     CONTINUE
        ENDIF
C
C ================================================
C *  COMPUTATION OF SYSTEM COEFFICIENT MATRICES  =
C ================================================
C
      DO 100 NN=1,NUMNP
C
        NC=NELE(NN,3)
        IF(NC.LT.O)THEN
          XCON=(X(NELE(NN,1))+X(NELE(NN,2)))/2.0
          YCON=(Y(NELE(NN,1))+Y(NELE(NN,2)))/2.0
        END IF
C
      DO 200 N=1,NUMELE
C
      N1=NELE(N,1)
      N2=NELE(N,2)
      N3=NELE(N,3)
      N4=NELE(N,4)
      N11=N1+NUMNP
      N22=N2+NUMNP
      N33=N3+NUMNP
      N44=N4+NUMNP
C
      IF(N3.LE.O)THEN
C
        IF(N3.LT.O)THEN
C
        CALL INTEG(XCON,YCON,N1,N2,N3,GI,GJ,HI,HJ,KI,KJ,LI,LJ,
```

171

```
     &                                  ITYPE,B1,B2,X,Y)
        ELSE
C
        CALL INTEG(X(NN),Y(NN),N1,N2,N3,GI,GJ,HI,HJ,KI,KJ,LI,LJ,
     &                                  ITYPE,B1,B2,X,Y)
        END IF
C
        GKERN(NN,N1)=GKERN(NN,N1)+GI
        GKERN(NN,N2)=GKERN(NN,N2)+GJ
        HKERN(NN,N1)=HKERN(NN,N1)+HI
        HKERN(NN,N2)=HKERN(NN,N2)+HJ
        GKERN(NN,N11)=GKERN(NN,N11)+KI
        GKERN(NN,N22)=GKERN(NN,N22)+KJ
        HKERN(NN,N11)=HKERN(NN,N11)+LI
        HKERN(NN,N22)=HKERN(NN,N22)+LJ
C
        HKERN(NN,NN)=HKERN(NN,NN)-HI-HJ
C
      ELSE
C
        IF(N4.EQ.O)THEN
C
          CALL INTEGQ(X,Y,X(NN),Y(NN),N1,N2,N3,GI,GJ,GK,HI,HJ,HK,
     $                KI,KJ,KK,LI,LJ,LK,ITYPE,B1,B2)
C
          GKERN(NN,N1)=GKERN(NN,N1)+GI
          GKERN(NN,N2)=GKERN(NN,N2)+GJ
          GKERN(NN,N3)=GKERN(NN,N3)+GK
          HKERN(NN,N1)=HKERN(NN,N1)+HI
          HKERN(NN,N2)=HKERN(NN,N2)+HJ
          HKERN(NN,N3)=HKERN(NN,N3)+HK
C
          GKERN(NN,N11)=GKERN(NN,N11)+KI
          GKERN(NN,N22)=GKERN(NN,N22)+KJ
          GKERN(NN,N33)=GKERN(NN,N33)+KK
          HKERN(NN,N11)=HKERN(NN,N11)+LI
          HKERN(NN,N22)=HKERN(NN,N22)+LJ
          HKERN(NN,N33)=HKERN(NN,N33)+LK
C
          HKERN(NN,NN)=HKERN(NN,NN)-HI-HJ-HK
C
        ELSE
C
          CALL INTEGO(X,Y,X(NN),Y(NN),N1,N2,N3,N4,GI,GJ,GK,GL,
     $                HI,HJ,HK,HL,KI,KJ,KK,KL,LI,LJ,LK,LL,ITYPE,B1,B2)
C
          GKERN(NN,N1)=GKERN(NN,N1)+GI
          GKERN(NN,N2)=GKERN(NN,N2)+GJ
          GKERN(NN,N3)=GKERN(NN,N3)+GK
          GKERN(NN,N4)=GKERN(NN,N4)+GL
C
          HKERN(NN,N1)=HKERN(NN,N1)+HI
          HKERN(NN,N2)=HKERN(NN,N2)+HJ
          HKERN(NN,N3)=HKERN(NN,N3)+HK
          HKERN(NN,N4)=HKERN(NN,N4)+HL
C
          GKERN(NN,N11)=GKERN(NN,N11)+KI
          GKERN(NN,N22)=GKERN(NN,N22)+KJ
          GKERN(NN,N33)=GKERN(NN,N33)+KK
          GKERN(NN,N44)=GKERN(NN,N44)+KL
C
          HKERN(NN,N11)=HKERN(NN,N11)+LI
          HKERN(NN,N22)=HKERN(NN,N22)+LJ
          HKERN(NN,N33)=HKERN(NN,N33)+LK
          HKERN(NN,N44)=HKERN(NN,N44)+LL
C
          HKERN(NN,NN)=HKERN(NN,NN)-HI-HJ-HK-HL
C
        END IF
```

```
C
      END IF
C
C     ==========================
C     =  COMPILE SOURCE TERM   =
C     ==========================
C
               B(NN)=B(NN)+B1
               B(NN+NUMNP)=B(NN+NUMNP)+B2
C
 200     CONTINUE
C
C     ===================================
C     =  POINT SOURCE TERM CALCULATION  =
C     ===================================
C
      IF(NSOUR.GE.1)THEN
C
         DO 210 IS=1,NSOUR
C
         IF(NC.LT.O)THEN
          DXS=SLX(IS)-XCON
          DYS=SLY(IS)-YCON
         ELSE
          DXS=SLX(IS)-X(NN)
          DYS=SLY(IS)-Y(NN)
         END IF
C
         ARG1=DXS*DXS+DYS*DYS
         ALNX=DLOG(ARG1)
C
         B(NN)=B(NN)-SVALUE(IS)*O.125*ARG1*(ALNX-2.0)
         B(NN+NUMNP)=B(NN+NUMNP)-SVALUE(IS)*O.5*ALNX
C
 210     CONTINUE
C
      END IF
C
         IF(ITYPE.EQ.3)THEN
            CALL FAN(X(NN),Y(NN),X,Y,NELE,B(NN),B(NN+NUMNP),Z1,Z2,Z3,Z4)
         END IF
C
 100     CONTINUE
C
         DO 300 I=1,NUMNP
          II=I+NUMNP
         DO 300 J=1,NUMNP
          JJ=J+NUMNP
           GKERN(II,JJ)=GKERN(I,J)
           HKERN(II,JJ)=HKERN(I,J)
 300     CONTINUE
C
C     ==========================================
C     =  ARRANGEMENT OF EQUATIONS FOR SOLUTION  =
C     ==========================================
C
      DO 400 J=1,NUMNP
       JJ=J+NUMNP
       IF(KODE(J).EQ.12)GOTO 410
       IF(KODE(J).EQ.13)GOTO 400
       IF(KODE(J).EQ.14)GOTO 420
       IF(KODE(J).EQ.23)GOTO 430
       IF(KODE(J).EQ.24)GOTO 440
C
      WRITE(IOUTDV,1000)
C
 410  DO 415 I=1,NUMEQ
       GIJ=GKERN(I,J)
       GKERN(I,J)=-HKERN(I,JJ)
```

```
 415      HKERN(I,JJ)=-GIJ
          GOTO 400
C
 420      DO 425 I=1,NUMEQ
          GIJ=GKERN(I,JJ)
          GKERN(I,JJ)=-HKERN(I,JJ)
 425      HKERN(I,JJ)=-GIJ
          GOTO 400
C
 430      DO 435 I=1,NUMEQ
          GIJ=GKERN(I,J)
          GKERN(I,J)=-HKERN(I,J)
 435      HKERN(I,J)=-GIJ
          GOTO 400
C
 440      DO 445 I=1,NUMEQ
          GIJ=GKERN(I,J)
          GKERN(I,J)=-HKERN(I,J)
          HKERN(I,J)=-GIJ
          GIJ=GKERN(I,JJ)
          GKERN(I,JJ)=-HKERN(I,JJ)
 445      HKERN(I,JJ)=-GIJ
C
 400  CONTINUE
C
C   ======================================
C   =  INITIALIZE THE SOLUTION VECTOR    =
C   ======================================
C
      DO 500 I=1,NUMEQ
        SOLV(I)=B(I)
      DO 500 J=1,NUMEQ
        SOLV(I)=SOLV(I)+HKERN(I,J)*FIXBND(J)
 500  CONTINUE
C
 1000 FORMAT (//,1X,' *** CODE NUMBER ERROR *** '//)
C
      RETURN
      END
```

6.5 Subroutine INTEG

INTEG performs all the necessary parameterized integrations of Equations (2.30) - (2.35) for a linear or a constant element. Depending on which of the two element types is used, the routine computes one or two components of the [G], [H], [K], and [L] matrices using the analytical expressions derived in Chapter 3. Also incorporated into this subroutine is an exact analysis for the evaluation of a special harmonic form of the source term f(x,y) and a numerical quadrature scheme for the evaluation of f(x,y) if the function satisfies the biharmonic equation, as developed in Chapter 4. Arbitrary source terms are analyzed by subroutine FAN using a domain "fanning" technique.

```
      SUBROUTINE INTEG(XP,YP,N1,N2,N3,GI,GJ,HI,HJ,KI,KJ,LI,LJ,
     $                               IB,B1,B2,X,Y)
C
C  ================================================================
C  =                                                              =
C  =   THIS SUBROUTINE INTEGRATES THE TERMS OF THE ASSEMBLY       =
C  =   MATRICES FOR LINEAR AND CONSTANT ELEMENTS                  =
C  =                                                              =
C  =   CALLING ARGUMENTS -                                        =
C  =                                                              =
C  =      XP,YP = COORDINATES OF SOURCE POINT                     =
C  =      XI,YI,                                                  =
C  =      XJ,YJ = COORDINATES OF ENDNODES OF ELEMENT BEING INTEGRATED =
C  =              IN CCW ORDER                                    =
C  =      HI,HJ = MATRIX TERMS OF 'HKERN' CORRESPONDING TO ENDNODE =
C  =              COORDINATES                                     =
C  =      GI,GJ = MATRIX TERMS OF 'GKERN' CORRESPONDING TO ENDNODE =
C  =              COORDINATES                                     =
C  =      KI,KJ = MATRIX TERMS OF 'KKERN' CORRESPONDING TO ENDNODE =
C  =              COORDINATES                                     =
C  =      LI,LJ = MATRIX TERMS OF 'LKERN' CORRESPONDING TO ENDNODE =
C  =              COORDINATES                                     =
C  =                                                              =
C  ================================================================
C
      IMPLICIT REAL*8(A-H,O-Z)
      REAL*8 KI,KJ,LI,LJ
      DIMENSION X(1),Y(1)
      COMMON/PCON/C1,C2,C3,C4
      COMMON/GQUAD/GK(100),GW(100),NUMGP
      COMMON/MODEL/NG1,NTOT,NTOT2,NUMNP,NUMELE,NUMEQ,NUMINT,ITYPE,IPLOT
      DATA PI/3.1415926535897932/
C
      B1=0.0
      B2=0.0
C
      XI=X(N1)
      XJ=X(N2)
      YI=Y(N1)
      YJ=Y(N2)
C
      AX=XJ-XI
      AY=YJ-YI
      BX=XI-XP
      BY=YI-YP
      CX=XJ-XP
      CY=YJ-YP
C
      A=AX*AX+AY*AY
      B=2*(AX*BX+AY*BY)
      C=BX*BX+BY*BY
      D=CX*CX+CY*CY
      CCO=4*A*C-B*B
      CC2=A+B+C
C
      ALEN=DSQRT(A)
      CLEN=DSQRT(C)
      DLEN=DSQRT(D)
      DCX=AY/ALEN
      DCY=-AX/ALEN
      E=BX*DCX+BY*DCY
C
      IF(CCO.LE.0.0)THEN
        CC1=1.0
        ANGLE=0.0
      ELSE
        CC1=DSQRT(CCO)
        ARG=(BX*CX+BY*CY)/(CLEN*DLEN)
        IF(ARG.GT.1.0)ARG=1.0
```

```
          IF(ARG.LT.-1.0)ARG=-1.0
          ANGLE=ACOS(ARG)
        END IF
C
      IF(C.EQ.0.0)THEN
        ALNC=0.0
      ELSE
        ALNC=DLOG(C)
      END IF
C
      IF(CC2.EQ.0.0)THEN
        ALNC2=0.0
      ELSE
        ALNC2=DLOG(CC2)
      END IF
C
      AIO=2.0*ANGLE/CC1
      AI1=(ALNC2-ALNC-B*AIO)/(2.0*A)
      AI2=(1.0-B*AI1-C*AIO)/A
      AI3=(0.5-B*AI2-C*AI1)/A
      AI4=((1.0/3.0)-B*AI3-C*AI2)/A
      AI5=(0.25-B*AI4-C*AI3)/A
C
      ALO=ALNC2-2.0*A*AI2-B*AI1
      AL1=(ALNC2-2.0*A*AI3-B*AI2)/2.0
      AL2=(ALNC2-2.0*A*AI4-B*AI3)/3.0
      AL3=(ALNC2-2.0*A*AI5-B*AI4)/4.0
C
      S1=A/3.0+B/2.0+C
      S2=A/4.0+B/3.0+C/2.0
C
      IF(N3.LT.0)THEN
C
        GI=-(ALEN/2.0)*ALO
        HI=-E*ALEN*AIO
        KI=-(ALEN/8.0)*(A*AL2+B*AL1+C*ALO-2.0*S1)
        LI=-E*(ALEN/4.0)*(ALO-1.0)
C
      ELSE
C
        GI=-(ALEN/2.0)*(ALO-AL1)
        GJ=-(ALEN/2.0)*AL1
        HI=-E*ALEN*(AIO-AI1)
        HJ=-E*ALEN*AI1
        KJ=(A*AL3+B*AL2+C*AL1-2.0*S2)
        KI=-(ALEN/8.0)*(A*AL2+B*AL1+C*ALO-2.0*S1-KJ)
        KJ=-(ALEN/8.0)*KJ
        LI=-E*(ALEN/4.0)*(ALO-AL1-0.5)
        LJ=-E*(ALEN/4.0)*(AL1-0.5)
C
      END IF
C
C ===========================================
C = HARMONIC SOURCE TERM OF FORM:           =
C =   F(X,Y) = C1* XY + C2 * X + C3 * Y + C4 =
C ===========================================
C
      IF(IB.EQ.1)THEN
C
      AI6=(0.20-B*AI5-C*AI4)/A
      AI7=((1.0/6.0)-B*AI6-C*AI5)/A
C
      AL4=(ALNC2-2.0*A*AI6-B*AI5)/5.0
      AL5=(ALNC2-2.0*A*AI7-B*AI6)/6.0
C
      S3=A/5.0+B/4.0+C/3.0
      P1=A*A
      P2=2*A*B
      P3=2*A*C+B*B
```

```
      P4=2*B*C
      P5=C*C
      T1=P1/5.O+P2/4.O+P3/3.O+P4/2.O+P5
      T2=P1/6.O+P2/5.O+P3/4.O+P4/3.O+P5/2.O
C
      G1=C1*AX*AY
      G2=C1*(AY*XI+AX*YI)+C2*AX+C3*AY
      G3=C1*XI*YI+C2*XI+C3*YI+C4
      H1=C1*(AY*DCX+AX*DCY)
      H2=(C1*YI+C2)*DCX+(C1*XI+C3)*DCY
C
      B1=(-ALEN/32.O)*(-H1*P1*AL5/4.O+(G1*E*A-(H1*P2+H2*P1)/4.O)*AL4
     &                +(E*(G1*B+G2*A)-(H1*P3+H2*P2)/4.O)*AL3
     &                +(E*(G1*C+G2*B+G3*A)-(H1*P4+H2*P3)/4.O)*AL2
     &                +(E*(G2*C+G3*B)-(H1*P5+H2*P4)/4.O)*AL1
     &                +(E*G3*C-(H2*P5/4.O))*ALO
     &                -(2.5*E)*(G1*S3+G2*S2+G3*S1)
     &                +O.75*(H1*T2+H2*T1))
C
      B2=(-ALEN/4.O)*((-H1*A/2.O)*AL3+(G1*E-(H1*B+H2*A)/2.O)*AL2
     &                +(G2*E-(H1*C+H2*B)/2.O)*AL1
     &                +(G3*E-(H2*C)/2.O)*ALO
     &                -E*(G1/3.O+G2/2.O+G3)+(H1*S2+H2*S1))
C
      END IF
C
C ================================================================
C =  GENERAL HARMONIC SOURCE TERM; DEFINE IN SUBROUTINE   =
C =  FUNCTION AND EVALUATED  BY GAUSSIAN QUADRATURE       =
C ================================================================
C
      IF(IB.EQ.2.OR.IB.EQ.1O)THEN
C
C ============================
C = BEGIN SURFACE INTEGRATION =
C ============================
C
      DO 1OO I=1,NUMGP
C
         GKI=GK(I)
         XG=AX*GKI+XI
         YG=AY*GKI+YI
         ARG1=A*GKI*GKI+B*GKI+C
         ALNX=DLOG(ARG1)
C
         B1F=(-F(XG,YG)*E/32.O)*ARG1*(ALNX-2.5)
         B1FN=(-FN(XG,YG,DCX,DCY)/128.O)*ARG1*ARG1*(ALNX-3.O)
         B1L=(-FL(XG,YG)*E/768.O)*ARG1*ARG1*(ALNX-1O.O/3.O)
         B1LN=(-FLN(XG,YG,DCX,DCY)/46O8.O)*ARG1**3*(ALNX-11.O/3.O)
C
         B2F=(-F(XG,YG)*E/4.O)*(ALNX-1.O)
         B2FN=(-FN(XG,YG,DCX,DCY)/8.O)*ARG1*(ALNX-2.O)
         B2L=(-FL(XG,YG)*E/32.O)*ARG1*(ALNX-2.5)
         B2LN=(-FLN(XG,YG,DCX,DCY)/128.O)*ARG1*ARG1*(ALNX-3.O)
C
         B1=B1+(B1F-B1FN+B1L-B1LN)*ALEN*GW(I)
         B2=B2+(B2F-B2FN+B2L-B2LN)*ALEN*GW(I)
C
 1OO     CONTINUE
C
      END IF
C
      RETURN
      END
```

6.6   Subroutine INTEGD

INTEGD is very similar to INTEG in both structure and capabilities. However, it is designed to evaluate the x and y derivative of both the field function and the Laplacian of the field function for a linear or a constant element. The analysis of the source term function f(x,y) in evaluating the x and y derivatives of the field variables has the same capabilities and limitations as described in subroutine INTEG.   To improve the computational efficiency, terms dealing with coordinate dependent variables were factored out of the formulation and replaced with more generalized variables.   The subroutine assigns the appropriate values to each of these generalized variables depending on the coordinate direction of the derivative desired.

```
      SUBROUTINE INTEGD(XP,YP,N1,N2,N3,GI,GJ,HI,HJ,KI,KJ,
     $                          LI,LJ,IB,B1,B2,ID,X,Y)
C
C ======================================================================
C =    THIS SUBROUTINE INTEGRATES THE TERMS INVOLVED WITH THE         =
C =    X AND Y-DERIVATIVES AT INTERNAL POINTS                         =
C ======================================================================
C
      IMPLICIT REAL*8(A-H,O-Z)
      REAL*8 KI,KJ,LI,LJ
      DIMENSION X(1),Y(1)
      COMMON/PCON/C1,C2,C3,C4
      COMMON/GQUAD/GK(100),GW(100),NUMGP
      COMMON/MODEL/NG1,NTOT,NTOT2,NUMNP,NUMELE,NUMEQ,NUMINT,ITYPE,IPLOT
      DATA PI/3.1415926535897932/
C
      B1=0.0
      B2=0.0
      GI=0.0
      GJ=0.0
      HI=0.0
      HJ=0.0
      KI=0.0
      KJ=0.0
      LI=0.0
      LJ=0.0
C
      XI=X(N1)
      XJ=X(N2)
      YI=Y(N1)
      YJ=Y(N2)
C
      AX=XJ-XI
      AY=YJ-YI
      BX=XI-XP
      BY=YI-YP
```

```
      CX=XJ-XP
      CY=YJ-YP
C
      A=AX*AX+AY*AY
      B=2*(AX*BX+AY*BY)
      C=BX*BX+BY*BY
      D=CX*CX+CY*CY
      CCO=4*A*C-B*B
      CC2=A+B+C
C
      ALEN=DSQRT(A)
      CLEN=DSQRT(C)
      DLEN=DSQRT(D)
      DCX=AY/ALEN
      DCY=-AX/ALEN
      E=BX*DCX+BY*DCY
C
C =========================================
C =  DECLARE X OR Y DERIVATIVE BASED ON ID  =
C =========================================
C
      IF(ID.EQ.1)THEN
         AD=AX
         BD=BX
         DRC=DCX
      ELSE
         AD=AY
         BD=BY
         DRC=DCY
      END IF
C
      IF(CCO.LE.0.0)THEN
         CC1=1.0
         ANGLE=0.0
      ELSE
         CC1=DSQRT(CCO)
         ARG=(BX*CX+BY*CY)/(CLEN*DLEN)
         IF(ARG.GT.1.0)ARG=1.0
         IF(ARG.LT.-1.0)ARG=-1.0
         ANGLE=ACOS(ARG)
      END IF
C
C =====================================
C =  TWO-DIMENSIONAL PLANAR ANALYSIS  =
C =====================================
C
      IF(C.EQ.0.0)THEN
         ALNC=0.0
      ELSE
         ALNC=DLOG(C)
      END IF
C
      IF(CC2.EQ.0.0)THEN
         ALNC2=0.0
      ELSE
         ALNC2=DLOG(CC2)
      END IF
C
      AIO=2.0*ANGLE/CC1
      AI1=(ALNC2-ALNC-B*AIO)/(2.0*A)
      AI2=(1.0-B*AI1-C*AIO)/A
      AI3=(0.5-B*AI2-C*AI1)/A
      AI4=((1.0/3.0)-B*AI3-C*AI2)/A
      AI5=(0.25-B*AI4-C*AI3)/A
C
      ALO=ALNC2-2.0*A*AI2-B*AI1
      AL1=(ALNC2-2.0*A*AI3-B*AI2)/2.0
      AL2=(ALNC2-2.0*A*AI4-B*AI3)/3.0
      AL3=(ALNC2-2.0*A*AI5-B*AI4)/4.0
```

```
C
      IF(CCO.EQ.0.0)THEN
        BA=B/(2.0*A)
        BA3=BA**3
        BD3=(1.0+BA)**3
        AS=A*A
        AI20=-(1.0/(3.0*BA*BD3)-1.0/(3.0*BA*BA3))/AS
        AI21=-1.0/(2.0*BD3*AS)+(BA/2.0)*AI20
      ELSE
        AI20=((2.0*A+B)/CC2-B/C+2.0*A*AIO)/CCO
        AI21=((2.0*C+B)/CC2-2.0+B*AIO)/(-CCO)
      END IF
        AI22=-1.0/(A*CC2)+(C/A)*AI20
        AI23=AI1/A-(C/A)*AI21-(B/A)*AI22
C
      IF(N3.LT.0)THEN
C
        GI=ALEN*(AD*AI1+BD*AIO)
        HI=ALEN*(DRC*AIO-2.0*E*(AD*AI21+BD*AI20))
        KI=(ALEN/4.0)*(AD*AL1+BD*ALO-(AD/2.0+BD))
        LI=(ALEN/4.0)*(2.0*E*(AD*AI1+BD*AIO)+DRC*(ALO-1.0))
C
      ELSE
C
        GJ=ALEN*(AD*AI2+BD*AI1)
        GI=ALEN*(AD*AI1+BD*AIO)-GJ
        HJ=ALEN*(DRC*AI1-2.0*E*(AD*AI22+BD*AI21))
        HI=ALEN*(DRC*AIO-2.0*E*(AD*AI21+BD*AI20))-HJ
        KJ=(ALEN/4.0)*(AD*AL2+BD*AL1-(AD/3.0+BD/2.0))
        KI=(ALEN/4.0)*(AD*AL1+BD*ALO-(AD/2.0+BD))-KJ
        LJ=(ALEN/4.0)*(2.0*E*(AD*AI2+BD*AI1)+DRC*(AL1-0.5))
        LI=(ALEN/4.0)*(2.0*E*(AD*AI1+BD*AIO)+DRC*(ALO-1.0))-LJ
C
      END IF
C
C     ===============================================
C     =   HARMONIC SOURCE TERM OF FORM:             =
C     =   F(X,Y) = C1* XY + C2 * X + C3 * Y + C4    =
C     ===============================================
C                   .
      IF(IB.EQ.1)THEN
C
      AI6=(0.20-B*AI5-C*AI4)/A
      AL4=(ALNC2-2.0*A*AI6-B*AI5)/5.0
C
      G1=C1*AX*AY
      G2=C1*(AY*XI+AX*YI)+C2*AX+C3*AY
      G3=C1*XI*YI+C2*XI+C3*YI+C4
      H1=C1*(AY*DCX+AX*DCY)
      H2=(C1*YI+C2)*DCX+(C1*XI+C3)*DCY
C
      P1=G1*AD
      P2=G1*BD+G2*AD
      P3=G2*BD+G3*AD
      P4=G3*BD
      Q1=H1*A*AD-A*G1*DRC
      Q2=H1*(A*BD+B*AD)+H2*A*AD-(A*G2+B*G1)*DRC
      Q3=H1*(B*BD+C*AD)+H2*(A*BD+B*AD)-(A*G3+B*G2+C*G1)*DRC
      Q4=H1*C*BD+H2*(B*BD+C*AD)-(B*G3+C*G2)*DRC
      Q5=H2*C*BD-C*G3*DRC
      R1=H1*AD
      R2=H1*BD+H2*AD
      R3=H2*BD
C
      B1=(ALEN/16.0)*(-Q1*AL4/2.0+(P1*E-Q2/2.0)*AL3+(P2*E-Q3/2.0)*AL2
     &                +(P3*E-Q4/2.0)*AL1+(P4*E-Q5/2.0)*ALO
     &                -1.5*E*(P1/4.0+P2/3.0+P3/2.0+P4)
     &                +1.25*(Q1/5.0+Q2/4.0+Q3/3.0+Q4/2.0+Q5))
C
```

```
      B2=(ALEN/4.0)*(2.0*E*(P1*AI3+P2*AI2+P3*AI1+P4*AIO)
     &                    +(G1*DRC-R1)*AL2+(G2*DRC-R2)*AL1+(G3*DRC-R3)*ALO
     &                    -DRC*(G1/3.0+G2/2.0+G3)+(R1/3.0+R2/2.0+R3))
C
      END IF
C
C   ============================================================
C   =  GENERAL HARMONIC SOURCE TERM; DEFINE IN SUBROUTINE      =
C   =  FUNCTION AND EVALUATED BY GAUSSIAN QUADRATURE           =
C   ============================================================
C
      IF(IB.EQ.2)THEN
C
C   =============================
C   = BEGIN SURFACE INTEGRATION =
C   =============================
C
      DO 100 I=1,NUMGP
C
         GKI=GK(I)
         XG=AX*GKI+XI
         YG=AY*GKI+YI
         ARG=AD*GKI+BD
         ARG1=A*GKI*GKI+B*GKI+C
         ALNX=DLOG(ARG1)
C
         B1F=(F(XG,YG)/32.0)*(2.0*E*ARG*(ALNX-1.5)
     &           +DRC*ARG1*(ALNX-2.5))
         B1FN=(FN(XG,YG,DCX,DCY)/32.0)*ARG1*(ALNX-2.5)*ARG
         B1L=(FL(XG,YG)/768.0)*(4.0*E*ARG*ARG1*(ALNX-17.0/6.0)
     &           +DRC*ARG1*ARG1*(ALNX-10.0/3.0))
         B1LN=(FLN(XG,YG,DCX,DCY)/768.0)*ARG1*ARG1*(ALNX-10.0/3.0)*ARG
C
         B2F=(F(XG,YG)/4.0)*(2*E*ARG/ARG1+DRC*(ALNX-1.0))
         B2FN=(FN(XG,YG,DCX,DCY)/4.0)*ARG*(ALNX-1.0)
         B2L=(FL(XG,YG)/32.0)*(2.0*E*ARG*(ALNX-1.5)
     &           +DRC*ARG1*(ALNX-2.5))
         B2LN=(FLN(XG,YG,DCX,DCY)/32.0)*ARG1*(ALNX-2.5)*ARG
C
         B1=B1+(B1F-B1FN+B1L-B1LN)*ALEN*GW(I)
         B2=B2+(B2F-B2FN+B2L-B2LN)*ALEN*GW(I)
C
 100     CONTINUE
C
      ENDIF
C
      RETURN
      END
```

6.7 Subroutine INTEGQ

INTEGQ evaluates the parameterized integrations defined in Equation (2.30) –
(2.35) for a quadratic element. Since a quadratic element is defined by three
boundary nodes as shown in Chapter 3, this routine computes three components of
the [G], [H], [K], and [L] matrices associated with that element. Due to the
possibility that each node in the element can be shared with an adjacent element,

some components may only be partially computed at this time. The remainder of each component is supplied by the integration over the adjacent elements.

In general a boundary node may be any one of the three nodes defining the quadratic element or any of the other points describing the domain surface. Therefore, the subroutine INTEGQ is formulated to handle each of the four possible cases and any resulting singularities in the integrations. Although this renders the total subroutine quite lengthy, the code itself is designed to execute only a small controlled segment at each use.

For a general curved geometry, the required integrations are evaluated numerically using the Gaussian points and weights generated by subroutine FORMGH. Also, the analysis is restricted to numerical evaluation of source terms $f(x,y)$ which satisfy the biharmonic equation. Arbitrary forms of $f(x,y)$ are evaluated using subroutine FAN.

A major improvement over existing formulations is that this subroutine has the intrinsic ability to sense rectilinear segments of the problem geometry. If such a configuration exists, those integrations over that segment are evaluated analytically by the subparametric expressions developed in Chapter 3 without loss of generality. In addition, if the entire boundary is composed of linear segments, the special harmonic form of $f(x,y)$, defined in Chapter 4, may be evaluated analytically.

A series of function statements labeled NI, NJ, NK, J, ARG, ARG1, and DC isolate the basic characteristics of the quadratic element formulation in parametric form. The first three NI, NJ, and NK define the shape functions, J is the Jacobian, ARG and ARG1 define two frequently encountered forms of the position vector, and finally, DC defines both the x and y direction cosines.

```
      SUBROUTINE INTEGQ(X,Y,XP,YP,N1,N2,N3,GI,GJ,GK,HI,HJ,HK,
     $                  KI,KJ,KK,LI,LJ,LK,IB,B1,B2)
C
C =================================================================
C =                                                               =
C =    THIS SUBROUTINE INTEGRATES THE TERMS OF THE ASSEMBLY       =
C =    MATRICES USING QUADRATIC  ELEMENTS                         =
C =                                                               =
C =    CALLING ARGUMENTS -                                        =
C =                                                               =
C =       XP,YP = COORDINATES OF SOURCE POINT                     =
C =       XI,YI,                                                  =
C =       XJ,YJ,                                                  =
C =       XK,YK = COORDINATES OF ENDNODES OF ELEMENT BEING INTEGRATED =
C =               IN CCW ORDER                                    =
C =    HI,HJ,HK = MATRIX TERMS OF 'HKERN' CORRESPONDING TO EACH NODE =
C =                                                               =
C =    GI,GJ,GK = MATRIX TERMS OF 'GKERN' CORRESPONDING TO EACH NODE =
C =                                                               =
C =    KI,KJ,KK = MATRIX TERMS OF 'KKERN' CORRESPONDING TO EACH NODE =
C =                                                               =
C =    LI,LJ,LK = MATRIX TERMS OF 'LKERN' CORRESPONDING TO EACH NODE =
C =                                                               =
C =================================================================
C
      IMPLICIT REAL*8(A-H,O-Z)
      REAL*8 KI,KJ,KK,LI,LJ,LK,NI,NJ,NK,J
      DIMENSION X(1),Y(1)
      COMMON/PCON/C1,C2,C3,C4
      COMMON/GQUAD/GP(100),GW(100),NUMGP
      COMMON/GLQUAD/GLP(7),GLW(7)
      COMMON/MODEL/NG1,NTOT,NTOT2,NUMNP,NUMELE,NUMEQ,NUMINT,ITYPE,IPLOT
      DATA PI/3.1415926535897932/
      DATA SMALL,SMALL1/1.0E-10,1.0E-2/
C
      LINEAR=0
      B1=0.0
      B2=0.0
      GI=0.0
      GJ=0.0
      GK=0.0
      HI=0.0
      HJ=0.0
      HK=0.0
      KI=0.0
      KJ=0.0
      KK=0.0
      LI=0.0
      LJ=0.0
      LK=0.0
C
      XI=X(N1)
      YI=Y(N1)
      XJ=X(N2)
      YJ=Y(N2)
      XK=X(N3)
      YK=Y(N3)
C
C ===============================
C = CHECK FOR LINEAR GEOMETRY =
C ===============================
C   .
      A1=XI-XJ
      A2=XK-XJ
      B1=YI-YJ
      B2=YK-YJ
      AREA=DABS(0.5*(A1*B2-B1*A2))
      IF(AREA.LE.SMALL)THEN
        ALEN1=A1*A1+B1*B1
```

```
       ALEN2=A2*A2+B2*B2
C         IF((ALEN1-ALEN2).LE.SMALL1)LINEAR=1
       END IF
C
       IF(LINEAR.EQ.O)THEN
C
C ======================================================
C = ANALYSIS FOR QUADRATIC GEOMETRY; INTEGRATIONS =
C = PERFORMED BY NUMERICAL QUADRATURE              =
C ======================================================
C
       AX=2.O*XI-4.O*XJ+2.O*XK
       AY=2.O*YI-4.O*YJ+2.O*YK
       BX=-3.O*XI+4.O*XJ-XK
       BY=-3.O*YI+4.O*YJ-YK
       DX=XI-XP
       DY=YI-YP
C
       A=AX*AX+AY*AY
       B=BX*BX+BY*BY
       C=2.O*(AX*BX+AY*BY)
       D=DX*DX+DY*DY
       E=2.O*(AX*DX+AY*DY)
       FF=2.O*(BX*DX+BY*DY)
C
C ====================================
C =  ANALYSIS FOR (XP,YP) = (XI,YI)  =
C ====================================
C
          XDF=DABS(XP-X(N1))
          YDF=DABS(YP-Y(N1))
          IF(XDF.LE.SMALL.AND.YDF.LE.SMALL)THEN
C
          DO 1O I=1,NUMGP
C
            GPI=GP(I)
            GWI=GW(I)
C
            AJ=J(GPI,A,B,C)
            XX=ARG(GPI,A,B,C,D,E,FF)
            XXT=A*GPI*GPI + C*GPI + B
            ALNX=DLOG(XX)
            ALNXT=DLOG(XXT)
            AXP=ARG1(GPI,AX,BX,DX)
            AYP=ARG1(GPI,AY,BY,DY)
            DCX=DC(GPI,AY,BY)
            DCY=-DC(GPI,AX,BX)
            DTERM=AXP*DCX+AYP*DCY
C
            C1=-O.5*ALNX*AJ*GWI
            C1T=-O.5*ALNXT*AJ*GWI
            C2=-(DTERM/XX)*GWI
            C3=-O.125*XX*(ALNX-2.O)*AJ*GWI
            C4=-O.25*(ALNX-1.O)*DTERM*GWI
C
            GI=GI+C1T*NI(GPI)
            GJ=GJ+C1*NJ(GPI)
            GK=GK+C1*NK(GPI)
C
            HJ=HJ+C2*NJ(GPI)
            HK=HK+C2*NK(GPI)
C
            KI=KI+C3*NI(GPI)
            KJ=KJ+C3*NJ(GPI)
            KK=KK+C3*NK(GPI)
C
            LI=LI+C4*NI(GPI)
            LJ=LJ+C4*NJ(GPI)
            LK=LK+C4*NK(GPI)
```

```
C
          IF(IB.EQ.2.OR.IB.EQ.10)THEN
C
            ADCX=DCX/AJ
            ADCY=DCY/AJ
            XG=NI(GPI)*XI+NJ(GPI)*XJ+NK(GPI)*XK
            YG=NI(GPI)*YI+NJ(GPI)*YJ+NK(GPI)*YK
C
        B1F=(-F(XG,YG)/32.0)*XX*(ALNX-2.5)*DTERM
        B1FN=(-FN(XG,YG,ADCX,ADCY)/128.0)*XX*XX*(ALNX-3.0)*AJ
        B1L=(-FL(XG,YG)/768.0)*XX*XX*(ALNX-10.0/3.0)*DTERM
        B1LN=(-FLN(XG,YG,ADCX,ADCY)/4608.0)*XX**3*(ALNX-11.0/3.0)*AJ
C
        B2F=(-F(XG,YG)/4.0)*(ALNX-1.0)*DTERM
        B2FN=(-FN(XG,YG,ADCX,ADCY)/8.0)*XX*(ALNX-2.0)*AJ
        B2L=(-FL(XG,YG)/32.0)*XX*(ALNX-2.5)*DTERM
        B2LN=(-FLN(XG,YG,ADCX,ADCY)/128.0)*XX*XX*(ALNX-3.0)*AJ
C
        B1=B1+(B1F-B1FN+B1L-B1LN)*GWI
        B2=B2+(B2F-B2FN+B2L-B2LN)*GWI
C
          END IF
C
 10       CONTINUE
C
          DO 15 I=1,7
C
          GLPI=GLP(I)
          GLWI=GLW(I)
C
          AJ=J(GLPI,A,B,C)
          C1=-AJ*GLWI
C
          GI=GI+NI(GLPI)*C1
C
 15       CONTINUE
C
          RETURN
          END IF
C
C ======================================
C =  ANALYSIS FOR (XP,YP) = (XJ,YJ)  =
C ======================================
C
          XDF=DABS(XP-X(N2))
          YDF=DABS(YP-Y(N2))
          IF(XDF.LE.SMALL.AND.YDF.LE.SMALL)THEN
C
          DO 20 I=1,NUMGP
C
          GPI=GP(I)
          GWI=GW(I)
          TS=0.5*(GPI+1.0)
          TU=0.5*(1.0-GPI)
C
          AJ=J(GPI,A,B,C)
          AJS=J(TS,A,B,C)
          AJU=J(TU,A,B,C)
C
          XXS=ARG(TS,A,B,C,D,E,FF)
          XXU=ARG(TU,A,B,C,D,E,FF)
C
          XXTS=(A/16.0)*GPI*GPI + (A/4.0 + C/8.0)*GPI
     &          + (3.0*(A+C)/8.0 + (B+E)/4.0)
          XXTU=(A/16.0)*GPI*GPI - (A/4.0 + C/8.0)*GPI
     &          + (3.0*(A+C)/8.0 + (B+E)/4.0)
C
          ALNXTS=DLOG(DABS(XXTS))
          ALNXTU=DLOG(DABS(XXTU))
```

```
            ALNXS=DLOG(XXS)
            ALNXU=DLOG(XXU)
C
            AXP=ARG1(GPI,AX,BX,DX)
            AYP=ARG1(GPI,AY,BY,DY)
            AXPS=ARG1(TS,AX,BX,DX)
            AYPS=ARG1(TS,AY,BY,DY)
            AXPU=ARG1(TU,AX,BX,DX)
            AYPU=ARG1(TU,AY,BY,DY)
C
            DCX=DC(GPI,AY,BY)
            DCY=-DC(GPI,AX,BX)
            DCXU=DC(TU,AY,BY)
            DCYU=-DC(TU,AX,BX)
            DCXS=DC(TS,AY,BY)
            DCYS=-DC(TS,AX,BX)
C
            DTERM=AXP*DCX+AYP*DCY
            DTERMS=AXPS*DCXS+AYPS*DCYS
            DTERMU=AXPU*DCXU+AYPU*DCYU
C
            C1ST=-0.25*ALNXTS*AJS*GWI
            C1UT=-0.25*ALNXTU*AJU*GWI
            C1S=-0.25*ALNXS*AJS*GWI
            C1U=-0.25*ALNXU*AJU*GWI
            C2S=-(0.5*DTERMS/XXS)*GWI
            C2U=-(0.5*DTERMU/XXU)*GWI
            C3S=-0.0625*XXS*(ALNXS-2.0)*AJS*GWI
            C3U=-0.0625*XXU*(ALNXU-2.0)*AJU*GWI
            C4S=-0.125*(ALNXS-1.0)*DTERMS*GWI
            C4U=-0.125*(ALNXU-1.0)*DTERMU*GWI
C
            GI=GI+(C1S*NI(TS)+C1U*NI(TU))
            GJ=GJ+(C1ST*NJ(TS)+C1UT*NJ(TU))
            GK=GK+(C1S*NK(TS)+C1U*NK(TU))
C
            HI=HI+(C2S*NI(TS)+C2U*NI(TU))
            HK=HK+(C2S*NK(TS)+C2U*NK(TU))
C
            KI=KI+(C3S*NI(TS)+C3U*NI(TU))
            KJ=KJ+(C3S*NJ(TS)+C3U*NJ(TU))
            KK=KK+(C3S*NK(TS)+C3U*NK(TU))
C
            LI=LI+(C4S*NI(TS)+C4U*NI(TU))
            LJ=LJ+(C4S*NJ(TS)+C4U*NJ(TU))
            LK=LK+(C4S*NK(TS)+C4U*NK(TU))
C
         IF(IB.EQ.2.OR.IB.EQ.10)THEN
C
            ADCXS=DCXS/AJS
            ADCYS=DCYS/AJS
            ADCXU=DCXU/AJU
            ADCYU=DCYU/AJU
C
            XGU=NI(TU)*XI+NJ(TU)*XJ+NK(TU)*XK
            YGU=NI(TU)*YI+NJ(TU)*YJ+NK(TU)*YK
            XGS=NI(TS)*XI+NJ(TS)*XJ+NK(TS)*XK
            YGS=NI(TS)*YI+NJ(TS)*YJ+NK(TS)*YK
C
            XG=NI(GPI)*XI+NJ(GPI)*XJ+NK(GPI)*XK
            YG=NI(GPI)*YI+NJ(GPI)*YJ+NK(GPI)*YK
            XX=ARG(GPI,A,B,C,D,E,FF)
            ALNX=DLOG(XX)
C
      B1F=(-F(XG,YG)/32.0)*XX*(ALNX-2.5)*DTERM
C
C     B1F=(-F(XGS,YGS)/64.0)*XXS*(ALNXS-2.5)*DTERMS
C    &    +(-F(XGU,YGU)/64.0)*XXU*(ALNXU-2.5)*DTERMU
      B1FN=(-FN(XGS,YGS,ADCXS,ADCYS)/256.0)*XXS*XXS*(ALNXS-3.0)*AJS
```

```
     &     +(-FN(XGU,YGU,ADCXU,ADCYU)/256.0)*XXU*XXU*(ALNXU-3.0)*AJU
      B1L=(-FL(XGS,YGS)/1536.0)*XXS*XXS*(ALNXS-10.0/3.0)*DTERMS
     &     +(-FL(XGU,YGU)/1536.0)*XXU*XXU*(ALNXU-10.0/3.0)*DTERMU
      B1LN=(-FLN(XGS,YGS,ADCXS,ADCYS)/9216.0)*XXS**3*(ALNXS-11.0/3.0)*AJS
     &     +(-FLN(XGU,YGU,ADCXU,ADCYU)/9216.0)*XXU**3*(ALNXU-11.0/3.0)*AJU
C
      B2F=(-F(XGS,YGS)/8.0)*(ALNXS-1.0)*DTERMS
     &     +(-F(XGU,YGU)/8.0)*(ALNXU-1.0)*DTERMU
      B2FN=(-FN(XGS,YGS,ADCXS,ADCYS)/16.0)*XXS*(ALNXS-2.0)*AJS
     &     +(-FN(XGU,YGU,ADCXU,ADCYU)/16.0)*XXU*(ALNXU-2.0)*AJU
      B2L=(-FL(XGS,YGS)/64.0)*XXS*(ALNXS-2.5)*DTERMS
     &     +(-FL(XGU,YGU)/64.0)*XXU*(ALNXU-2.5)*DTERMU
      B2LN=(-FLN(XGS,YGS,ADCXS,ADCYS)/256.0)*XXS*XXS*(ALNXS-3.0)*AJS
     &     +(-FLN(XGU,YGU,ADCXU,ADCYU)/256.0)*XXU*XXU*(ALNXU-3.0)*AJU
C
         B1=B1+(B1F-B1FN+B1L-B1LN)*GWI
         B2=B2+(B2F-B2FN+B2L-B2LN)*GWI
C
            END IF
C
 20         CONTINUE
C
            DO 25 I=1,7
C
            GLPI=GLP(I)
            GLWI=GLW(I)
C
            TS=0.5*(GLPI+1.0)
            TU=0.5*(1.0-GLPI)
C
            AJS=J(TS,A,B,C)
            AJU=J(TU,A,B,C)
C
            C1S=-0.5*AJS*GLWI
            C1U=-0.5*AJU*GLWI
C
            GJ=GJ+(C1S*NJ(TS)+C1U*NJ(TU))
C
 25         CONTINUE
C
            RETURN
            END IF
C
C ===================================
C =  ANALYSIS FOR (XP,YP) = (XK,YK)  =
C ===================================
C
            XDF=DABS(XP-X(N3))
            YDF=DABS(YP-Y(N3))
            IF(XDF.LE.SMALL.AND.YDF.LE.SMALL)THEN
C
            DO 30 I=1,NUMGP
C
            GPI=GP(I)
            GWI=GW(I)
C
            TS=1.0-GPI
            AJ=J(TS,A,B,C)
            XX=ARG(TS,A,B,C,D,E,FF)
            XXT=A*GPI*GPI-(4.0*A + C)*GPI+(6.0*A + 3.0*C + (B+E))
            ALNX=DLOG(XX)
            ALNXT=DLOG(XXT)
            AXP=ARG1(TS,AX,BX,DX)
            AYP=ARG1(TS,AY,BY,DY)
            DCX=DC(TS,AY,BY)
            DCY=-DC(TS,AX,BX)
            DTERM=AXP*DCX+AYP*DCY
C
            C1=-0.5*ALNX*AJ*GWI
```

```
            C1T=-0.5*ALNXT*AJ*GWI
            C2=-(DTERM/XX)*GWI
            C3=-0.125*XX*(ALNX-2.0)*AJ*GWI
            C4=-0.25*(ALNX-1.0)*DTERM*GWI
C
            GI=GI+C1*NI(TS)
            GJ=GJ+C1*NJ(TS)
            GK=GK+C1T*NK(TS)
C
            HI=HI+C2*NI(TS)
            HJ=HJ+C2*NJ(TS)
C
            KI=KI+C3*NI(TS)
            KJ=KJ+C3*NJ(TS)
            KK=KK+C3*NK(TS)
C
            LI=LI+C4*NI(TS)
            LJ=LJ+C4*NJ(TS)
            LK=LK+C4*NK(TS)
C
        IF(IB.EQ.2.OR.IB.EQ.10)THEN
C
            ADCX=DCX/AJ
            ADCY=DCY/AJ
C
            XGS=NI(TS)*XI+NJ(TS)*XJ+NK(TS)*XK
            YGS=NI(TS)*YI+NJ(TS)*YJ+NK(TS)*YK
C
        B1F=(-F(XGS,YGS)/32.0)*XX*(ALNX-2.5)*DTERM
        B1FN=(-FN(XGS,YGS,ADCX,ADCY)/128.0)*XX*XX*(ALNX-3.0)*AJ
        B1L=(-FL(XG,YG)/768.0)*XX*XX*(ALNX-10.0/3.0)*DTERM
        B1LN=(-FLN(XGS,YGS,ADCX,ADCY)/4608.0)*XX**3*(ALNX-11.0/3.0)*AJ
C
        B2F=(-F(XGS,YGS)/4.0)*(ALNX-1.0)*DTERM
        B2FN=(-FN(XGS,YGS,ADCX,ADCY)/8.0)*XX*(ALNX-2.0)*AJ
        B2L=(-FL(XGS,YGS)/32.0)*XX*(ALNX-2.5)*DTERM
        B2LN=(-FLN(XGS,YGS,ADCX,ADCY)/128.0)*XX*XX*(ALNX-3.0)*AJ
C
        B1=B1+(B1F-B1FN+B1L-B1LN)*GWI
        B2=B2+(B2F-B2FN+B2L-B2LN)*GWI
C
            END IF
C
C
  30        CONTINUE
C
            DO 35 I=1,7
C
            GLPI=GLP(I)
            GLWI=GLW(I)
            TS=1.0-GLPI
C
            AJ=J(TS,A,B,C)
            C1=-AJ*GLWI
C
            GK=GK+NK(TS)*C1
C
  35        CONTINUE
C
          RETURN
          END IF
C
          DO 40 I=1,NUMGP
C
            GPI=GP(I)
            GWI=GW(I)
C
            AJ=J(GPI,A,B,C)
            XX=ARG(GPI,A,B,C,D,E,FF)
            ALNX=DLOG(XX)
```

```
              AXP=ARG1(GPI,AX,BX,DX)
              AYP=ARG1(GPI,AY,BY,DY)
              DCX=DC(GPI,AY,BY)
              DCY=-DC(GPI,AX,BX)
              DTERM=AXP*DCX+AYP*DCY
C
              C1=-0.5*ALNX*AJ*GWI
              C2=-(DTERM/XX)*GWI
              C3=-0.125*XX*(ALNX-2.0)*AJ*GWI
              C4=-0.25*(ALNX-1.0)*DTERM*GWI
C
              GI=GI+C1*NI(GPI)
              GJ=GJ+C1*NJ(GPI)
              GK=GK+C1*NK(GPI)
C
              HI=HI+C2*NI(GPI)
              HJ=HJ+C2*NJ(GPI)
              HK=HK+C2*NK(GPI)
C
              KI=KI+C3*NI(GPI)
              KJ=KJ+C3*NJ(GPI)
              KK=KK+C3*NK(GPI)
C
              LI=LI+C4*NI(GPI)
              LJ=LJ+C4*NJ(GPI)
              LK=LK+C4*NK(GPI)
C
             IF(IB.EQ.2.OR.IB.EQ.10)THEN
C
              XG=NI(GPI)*XI+NJ(GPI)*XJ+NK(GPI)*XK
              YG=NI(GPI)*YI+NJ(GPI)*YJ+NK(GPI)*YK
              ADCX=DCX/AJ
              ADCY=DCY/AJ
C
          B1F=(-F(XG,YG)/32.0)*XX*(ALNX-2.5)*DTERM
          B1FN=(-FN(XG,YG,ADCX,ADCY)/128.0)*XX*XX*(ALNX-3.0)*AJ
          B1L=(-FL(XG,YG)/768.0)*XX*XX*(ALNX-10.0/3.0)*DTERM
          B1LN=(-FLN(XG,YG,ADCX,ADCY)/4608.0)*XX**3*(ALNX-11.0/3.0)*AJ
C
          B2F=(-F(XG,YG)/4.0)*(ALNX-1.0)*DTERM
          B2FN=(-FN(XG,YG,ADCX,ADCY)/8.0)*XX*(ALNX-2.0)*AJ
          B2L=(-FL(XG,YG)/32.0)*XX*(ALNX-2.5)*DTERM
          B2LN=(-FLN(XG,YG,ADCX,ADCY)/128.0)*XX*XX*(ALNX-3.0)*AJ
C
          B1=B1+(B1F-B1FN+B1L-B1LN)*GWI
          B2=B2+(B2F-B2FN+B2L-B2LN)*GWI
C
             END IF
C
 40       CONTINUE
C
          RETURN
C
          ELSE
C
C ==========================================
C = QUADRATIC ANALYSIS FOR LINEAR GEOMETRY; =
C = EXACT INTEGRATIONS                      =
C ==========================================
C
      AX=XK-XI
      AY=YK-YI
      BX=XI-XP
      BY=YI-YP
      CX=XK-XP
      CY=YK-YP
C
      A=AX*AX+AY*AY
      B=2*(AX*BX+AY*BY)
```

```
      C=BX*BX+BY*BY
      D=OX*CX+CY*CY
      CCO=4*A*C-B*B
      CC2=A+B+C
C
      ALEN=DSQRT(A)
      CLEN=DSQRT(C)
      DLEN=DSQRT(D)
      DCX=AY/ALEN
      DCY=-AX/ALEN
      E=BX*DCX+BY*DCY
C
      IF(CCO.LE.O.O)THEN
        CC1=1.O
        ANGLE=O.O
      ELSE
        CC1=DSQRT(CCO)
        AARG=(BX*CX+BY*CY)/(CLEN*DLEN)
        IF(AARG.GT.1.O)AARG=1.O
        IF(AARG.LT.-1.O)AARG=-1.O
        ANGLE=ACOS(AARG)
      END IF
C
      IF(C.EQ.O.O)THEN
        ALNC=O.O
      ELSE
        ALNC=DLOG(C)
      END IF
C
      IF(CC2.EQ.O.O)THEN
        ALNC2=O.O
      ELSE
        ALNC2=DLOG(CC2)
      END IF
C
      AIO=2.O*ANGLE/CC1
      AI1=(ALNC2-ALNC-B*AIO)/(2.O*A)
      AI2=(1.O-B*AI1-C*AIO)/A
      AI3=(O.5-B*AI2-C*AI1)/A
      AI4=((1.O/3.O)-B*AI3-C*AI2)/A
      AI5=(O.25-B*AI4-C*AI3)/A
      AI6=(O.2O-B*AI5-C*AI4)/A
C
      ALO=ALNC2-2.O*A*AI2-B*AI1
      AL1=(ALNC2-2.O*A*AI3-B*AI2)/2.O
      AL2=(ALNC2-2.O*A*AI4-B*AI3)/3.O
      AL3=(ALNC2-2.O*A*AI5-B*AI4)/4.O
      AL4=(ALNC2-2.O*A*AI6-B*AI5)/5.O
C
      GI=-(ALEN/2.O)*(2.O*AL2-3.O*AL1+ALO)
      GJ=-ALEN*2.O*(-AL2+AL1)
      GK=-(ALEN/2.O)*(2.O*AL2-AL1)
      HI=-E*ALEN*(2.O*AI2-3.O*AI1+AIO)
      HJ=-E*ALEN*4.O*(-AI2+AI1)
      HK=-E*ALEN*(2.O*AI2-AI1)
      KI=-(ALEN/8.O)*(2.O*A*AL4+(2.O*B-3.O*A)*AL3+(A-3.O*B+2.O*C)*AL2
     &                +(B-3.O*C)*AL1+C*ALO-2.O*(2.O*A/5.O+(2.O*B-3.O*A)/4.O
     &                +(A-3.O*B+2.O*C)/3.O+(B-3.O*C)/2+C))
      KJ=-(ALEN/2.O)*(-A*AL4+(A-B)*AL3+(B-C)*AL2+C*AL1-2.O*(-A/5.O
     &                +(A-B)/4.O+(B-C)/3.O+C/2.O))
      KK=-(ALEN/8.O)*(2.O*A*AL4+(2.O*B-A)*AL3+(2.O*C-B)*AL2-C*AL1
     &                -2.O*(2.O*A/5.O+(2.O*B-A)/4.O+(2.O*C-B)/3.O-C/2.O))
      LI=-E*(ALEN/4.O)*(2.O*AL2-3.O*AL1+ALO-1.O/6.O)
      LJ=-E*ALEN*(-AL2+AL1-1.O/6.O)
      LK=-E*(ALEN/4.O)*(2.O*AL2-AL1-1.O/6.O)
C
C =============================================
C =  HARMONIC SOURCE TERM OF FORM:             =
C =    F(X,Y) = C1* XY + C2 * X + C3 * Y + C4  =
```

```fortran
C     ==========================================
C
      IF(IB.EQ.1)THEN
C
      AI7=((1.0/6.0)-B*AI6-C*AI5)/A
      AL5=(ALNC2-2.0*A*AI7-B*AI6)/6.0
C
      S1=A/3.0+B/2.0+C
      S2=A/4.0+B/3.0+C/2.0
      S3=A/5.0+B/4.0+C/3.0
      P1=A*A
      P2=2*A*B
      P3=2*A*C+B*B
      P4=2*B*C
      P5=C*C
      T1=P1/5.0+P2/4.0+P3/3.0+P4/2.0+P5
      T2=P1/6.0+P2/5.0+P3/4.0+P4/3.0+P5/2.0
C
      G1=C1*AX*AY
      G2=C1*(AY*XI+AX*YI)+C2*AX+C3*AY
      G3=C1*XI*YI+C2*XI+C3*YI+C4
      H1=C1*(AY*DCX+AX*DCY)
      H2=(C1*YI+C2)*DCX+(C1*XI+C3)*DCY
C
      B1=(-ALEN/32.0)*(-H1*P1*AL5/4.0+(G1*E*A-(H1*P2+H2*P1)/4.0)*AL4
     &              +(E*(G1*B+G2*A)-(H1*P3+H2*P2)/4.0)*AL3
     &              +(E*(G1*C+G2*B+G3*A)-(H1*P4+H2*P3)/4.0)*AL2
     &              +(E*(G2*C+G3*B)-(H1*P5+H2*P4)/4.0)*AL1
     &              +(E*G3*C-(H2*P5/4.0))*ALO
     &              -(2.5*E)*(G1*S3+G2*S2+G3*S1)
     &              +0.75*(H1*T2+H2*T1))
C
      B2=(-ALEN/4.0)*((-H1*A/2.0)*AL3+(G1*E-(H1*B+H2*A)/2.0)*AL2
     &              +(G2*E-(H1*C+H2*B)/2.0)*AL1
     &              +(G3*E-(H2*C)/2.0)*ALO
     &              -E*(G1/3.0+G2/2.0+G3)+(H1*S2+H2*S1))
C
      END IF
C
C     ============================================================
C   = GENERAL BIHARMONIC SOURCE TERM; DEFINED IN SUBROUTINE      =
C   = FUNCTION AND EVALUATED BY GAUSSIAN QUADRATURE              =
C     ============================================================
C
      IF(IB.EQ.2.OR.IB.EQ.10)THEN
C
      DO 100 I=1,NUMGP
C
         GKI=GP(I)
         XG=AX*GKI+XI
         YG=AY*GKI+YI
         AARG1=A*GKI*GKI+B*GKI+C
         ALNX=DLOG(AARG1)
C
         B1F=(-F(XG,YG)*E/32.0)*AARG1*(ALNX-2.5)
         B1FN=(-FN(XG,YG,DCX,DCY)/128.0)*AARG1*AARG1*(ALNX-3.0)
         B1L=(-FL(XG,YG)*E/768.0)*AARG1*AARG1*(ALNX-10.0/3.0)
         B1LN=(-FLN(XG,YG,DCX,DCY)/4608.0)*AARG1**3*(ALNX-11.0/3.0)
C
         B2F=(-F(XG,YG)*E/4.0)*(ALNX-1.0)
         B2FN=(-FN(XG,YG,DCX,DCY)/8.0)*AARG1*(ALNX-2.0)
         B2L=(-FL(XG,YG)*E/32.0)*AARG1*(ALNX-2.5)
         B2LN=(-FLN(XG,YG,DCX,DCY)/128.0)*AARG1*AARG1*(ALNX-3.0)
C
         B1=B1+(B1F-B1FN+B1L-B1LN)*ALEN*GW(I)
         B2=B2+(B2F-B2FN+B2L-B2LN)*ALEN*GW(I)
C
 100     CONTINUE
         END IF
```

```
C
        END IF
        RETURN
        END
C
C----------------------------------------------- NI SHAPE FUNCTION
C
        FUNCTION NI(T)
        IMPLICIT REAL*8(A-H,O-Z)
        REAL*8 NI
C
          NI = 2.0*T*T - 3.0*T + 1.0
C
        RETURN
        END
C
C----------------------------------------------- NJ SHAPE FUNCTION
C
        FUNCTION NJ(T)
        IMPLICIT REAL*8(A-H,O-Z)
        REAL*8 NJ
C
          NJ = 4.0*T*(1-T)
C
        RETURN
        END
C
C----------------------------------------------- NK SHAPE FUNCTION
C
        FUNCTION NK(T)
        IMPLICIT REAL*8(A-H,O-Z)
        REAL *8 NK
C
          NK = 2.0*T*T - T
C
        RETURN
        END
C
C----------------------------------------------- JACOBIAN
C
        FUNCTION J(T,A,B,C)
        IMPLICIT REAL*8(A-H,O-Z)
        REAL*8 J
C
          J = DSQRT(4.0*A*T*T + 2.0*C*T + B)
C
        RETURN
        END
C
C----------------------------------------------- FUNCTION ARG
C
        FUNCTION ARG(T,A,B,C,D,E,FF)
        IMPLICIT REAL*8(A-H,O-Z)
C
          TS=T*T
          ARG = A*(TS*TS) + C*(TS*T) + (B+E)*TS + FF*T + D
C
        RETURN
        END
C
C----------------------------------------------- FUNCTION ARG1
C
        FUNCTION ARG1(T,A,B,C)
        IMPLICIT REAL*8(A-H,O-Z)
C
          ARG1 = A*T*T + B*T + C
C
        RETURN
        END
```

```
C
C------------------------------------- DIRECTION COSINE FUNCTION
C
      FUNCTION DC(T,A,B)
      IMPLICIT REAL*8(A-H,O-Z)
C
        DC = 2.0*A*T + B
C
      RETURN
      END
C
C-------------------------------------------------- BLOCK DATA GLQUAD
C
      BLOCK DATA GLQPTS
C
      REAL*8 GLP,GLW
      COMMON/GLQUAD/GLP(7),GLW(7)
C
      DATA GLP/0.01671935,0.10018567,0.24629424,0.43346349,
     &         0.63235098,0.81111862,0.94084816/
C
      DATA GLW/-0.19616938,-0.27030264,-0.23968187,-0.16577577,
     &         -0.08894322,-0.03319430,-0.00593278/
C
      END
```

6.8 Subroutine INTGQD

INTGQD has a structure much like its linear/constant element counterpart INTEGD, except that it is written for quadratic elements. The subroutine is used in the computation of components of the x and y derivatives of the field function and its Laplacian at arbitrary internal point locations in the problem domain. For a general geometry, the parameterized integrations involving the boundary quantities and any possible biharmonic form of the source term f(x,y) are determined numerically. However, if the boundary is segmentally linear the special harmonic form of f(x,y) may be evaluated using an exact analysis similar to that used in subroutine INTEGD.

```
      SUBROUTINE INTGQD(X,Y,XP,YP,N1,N2,N3,GI,GJ,GK,HI,HJ,HK,
     $                  KI,KJ,KK,LI,LJ,LK,IB,B1,B2,ID)
C
C  =================================================================
C  =    THIS SUBROUTINE INTEGRATES THE TERMS INVOLVED WITH THE      =
C  =    X-DERIVATIVES AT INTERNAL POINTS USING QUADRATIC  ELEMENTS  =
C  =================================================================
C
      IMPLICIT REAL*8(A-H,O-Z)
      REAL*8 KI,KJ,KK,KL,LI,LJ,LK,LL,NI,NJ,NK,J
      DIMENSION X(1),Y(1)
      COMMON/PCON/C1,C2,C3,C4
      COMMON/GQUAD/GP(100),GW(100),NUMGP
      COMMON/MODEL/NG1,NTOT,NTOT2,NUMNP,NUMELE,NUMEQ,NUMINT,ITYPE,IPLOT
      DATA PI/3.1415926535897932/
      DATA SMALL,SMALL1/1.0E-10,1.0E-2/
C
      LINEAR=O
      B1=0.0
      B2=0.0
      GI=0.0
      GJ=0.0
      GK=0.0
      HI=0.0
      HJ=0.0
      HK=0.0
      KI=0.0
      KJ=0.0
      KK=0.0
      LI=0.0
      LJ=0.0
      LK=0.0
C
      XI=X(N1)
      YI=Y(N1)
      XJ=X(N2)
      YJ=Y(N2)
      XK=X(N3)
      YK=Y(N3)
C
C ------------------------------
C = CHECK FOR LINEAR GEOMETRY =
C ==============================
C
      AC1=XI-XJ
      AC2=XK-XJ
      BC1=YI-YJ
      BC2=YK-YJ
      AREA=DABS(0.5*(AC1*BC2-BC1*AC2))
       IF(AREA.LE.SMALL)THEN
         ALEN1=AC1*AC1+BC1*BC1
         ALEN2=AC2*AC2+BC2*BC2
C         IF((ALEN1-ALEN2).LE.SMALL1)LINEAR=1
       END IF
C
      IF(LINEAR.EQ.O)THEN
C
C ===================================================
C = ANALYSIS FOR QUADRATIC GEOMETRY; INTEGRATIONS =
C = PERFORMED BY NUMERICAL QUADRATURE             =
C ===================================================
C
      AX=2.0*XI-4.0*XJ+2.0*XK
      AY=2.0*YI-4.0*YJ+2.0*YK
      BX=-3.0*XI+4.0*XJ-XK
      BY=-3.0*YI+4.0*YJ-YK
      DX=XI-XP
      DY=YI-YP
```

```
C
      A=AX*AX+AY*AY
      B=BX*BX+BY*BY
      C=2.O*(AX*BX+AY*BY)
      D=DX*DX+DY*DY
      E=2.O*(AX*DX+AY*DY)
      FF=2.O*(BX*DX+BY*DY)
C
C ============================
C =  BEGIN SURFACE QUADRATURE =
C ============================
C
      DO 40 I=1,NUMGP
C
         GPI=GP(I)
         GWI=GW(I)
C
         AJ=J(GPI,A,B,C)
         XX=ARG(GPI,A,B,C,D,E,FF)
         ALNX=DLOG(XX)
         AXP=ARG1(GPI,AX,BX,DX)
         AYP=ARG1(GPI,AY,BY,DY)
         DCX=DC(GPI,AY,BY)/AJ
         DCY=-DC(GPI,AX,BX)/AJ
         DTERM=AXP*DCX+AYP*DCY
C
C ===============================================
C =  DECLARE X OR Y DERIVATIVE BASED ON ID  =
C ===============================================
C
      IF(ID.EQ.1)THEN
        AD=AXP
        DRC=DCX
      ELSE
        AD=AYP
        DRC=DCY
      END IF
C
         C1=(AD/XX)*AJ*GWI
         C2=(DRC/XX-2.O*AD*DTERM/(XX*XX))*AJ*GWI
         C3=O.25*AD*(ALNX-1.O)*AJ*GWI
         C4=O.25*(2.O*AD*DTERM/XX+DRC*(ALNX-1.O))*AJ*GWI
C
         GI=GI+C1*NI(GPI)
         GJ=GJ+C1*NJ(GPI)
         GK=GK+C1*NK(GPI)
C
         HI=HI+C2*NI(GPI)
         HJ=HJ+C2*NJ(GPI)
         HK=HK+C2*NK(GPI)
C
         KI=KI+C3*NI(GPI)
         KJ=KJ+C3*NJ(GPI)
         KK=KK+C3*NK(GPI)
C
         LI=LI+C4*NI(GPI)
         LJ=LJ+C4*NJ(GPI)
         LK=LK+C4*NK(GPI)
C
C ==================================================================
C = GENERAL BIHARMONIC SOURCE TERM; DEFINED IN SUBROUTINE   =
C = FUNCTION AND EVALUATED BY GAUSSIAN QUADRATURE           =
C ==================================================================
C
         IF(IB.EQ.2.OR.IB.EQ.1O)THEN
C
            XG=NI(GPI)*XI+NJ(GPI)*XJ+NK(GPI)*XK
            YG=NI(GPI)*YI+NJ(GPI)*YJ+NK(GPI)*YK
```

```
C
        B1F=(F(XG,YG)/32.O)*(2.O*DTERM*AD*(ALNX-1.5)
     &        +DRC*XX*(ALNX-2.5))
        B1FN=(FN(XG,YG,DCX,DCY)/32.O)*XX*(ALNX-2.5)*AD
        B1L=(FL(XG,YG)/768.O)*(4.O*DTERM*AD*XX*(ALNX-17.O/6.O)
     &        +DRC*XX*XX*(ALNX-10.O/3.O))
        B1LN=(FLN(XG,YG,DCX,DCY)/768.O)*XX*XX*(ALNX-10.O/3.O)*AD
C
        B2F=(F(XG,YG)/4.O)*(2.O*DTERM*AD/XX+DRC*(ALNX-1.O))
        B2FN=(FN(XG,YG,DCX,DCY)/4.O)*AD*(ALNX-1.O)
        B2L=(FL(XG,YG)/32.O)*(2.O*DTERM*AD*(ALNX-1.5)
     &        +DRC*XX*(ALNX-2.5))
        B2LN=(FLN(XG,YG,DCX,DCY)/32.O)*XX*(ALNX-2.5)*AD
C
        B1=B1+(B1F-B1FN+B1L-B1LN)*AJ*GWI
        B2=B2+(B2F-B2FN+B2L-B2LN)*AJ*GWI
C
           END IF
C
  40      CONTINUE
        RETURN
C
        ELSE
C
C ==========================================
C = QUADRATIC ANALYSIS FOR LINEAR GEOMETRY; =
C = EXACT INTEGRATIONS                      =
C ==========================================
C
        AX=XK-XI
        AY=YK-YI
        BX=XI-XP
        BY=YI-YP
        CX=XK-XP
        CY=YK-YP
C
        A=AX*AX+AY*AY
        B=2*(AX*BX+AY*BY)
        C=BX*BX+BY*BY
        D=CX*CX+CY*CY
        CCO=4*A*C-B*B
        CC2=A+B+C
C
        ALEN=DSQRT(A)
        CLEN=DSQRT(C)
        DLEN=DSQRT(D)
        DCX=AY/ALEN
        DCY=-AX/ALEN
        E=BX*DCX+BY*DCY
C
C ==========================================
C =  DECLARE X OR Y DERIVATIVE BASED ON ID  =
C ==========================================
C
        IF(ID.EQ.1)THEN
          AD=AX
          BD=BX
          DRC=DCX
        ELSE
          AD=AY
          BD=BY
          DRC=DCY
        END IF
C
        IF(CCO.LE.O.O)THEN
          CC1=1.O
          ANGLE=O.O
        ELSE
```

```
            CC1=DSQRT(CCO)
            AARG=(BX*CX+BY*CY)/(CLEN*DLEN)
            IF(AARG.GT.1.0)AARG=1.0
            IF(AARG.LT.-1.0)AARG=-1.0
            ANGLE=ACOS(AARG)
          END IF
C
          IF(C.EQ.0.0)THEN
            ALNC=0.0
          ELSE
            ALNC=DLOG(C)
          END IF
C
          IF(CC2.EQ.0.0)THEN
            ALNC2=0.0
          ELSE
            ALNC2=DLOG(CC2)
          END IF
C
          AIO=2.0*ANGLE/CC1
          AI1=(ALNC2-ALNC-B*AIO)/(2.0*A)
          AI2=(1.0-B*AI1-C*AIO)/A
          AI3=(0.5-B*AI2-C*AI1)/A
          AI4=((1.0/3.0)-B*AI3-C*AI2)/A
          AI5=(0.25-B*AI4-C*AI3)/A
C
          ALO=ALNC2-2.0*A*AI2-B*AI1
          AL1=(ALNC2-2.0*A*AI3-B*AI2)/2.0
          AL2=(ALNC2-2.0*A*AI4-B*AI3)/3.0
          AL3=(ALNC2-2.0*A*AI5-B*AI4)/4.0
C
          IF(CCO.EQ.0.0)THEN
            BA=B/(2.0*A)
            BA3=BA**3
            BD3=(1.0+BA)**3
            AS=A*A
            AI2O=-(1.0/(3.0*BA*BD3)-1.0/(3.0*BA*BA3))/AS
            AI21=-1.0/(2.0*BD3*AS)+(BA/2.0)*AI2O
C            AI22=-1.0/(BD3*AS)+2.0*BA*AI21
          ELSE
            AI2O=((2.0*A+B)/CC2-B/C+2.0*A*AIO)/CCO
            AI21=((2.0*C+B)/CC2-2.0*B*AIO)/(-CCO)
C            AI22=((B*B-2.0*A*C+B*C)/(A*CC2)-B/A+2.0*C*AIO)/CCO
          END IF
            AI22=-1.0/(A*CC2)+(C/A)*AI2O
            AI23=AI1/A-(C/A)*AI21-(B/A)*AI22
C
          GI=ALEN*(2.0*AD*AI3+(2.0*BD-3.0*AD)*AI2+(AD-3.0*BD)*AI1+BD*AIO)
          GJ=ALEN*4.0*(-AD*AI3+(AD-BD)*AI2+BD*AI1)
          GK=ALEN*(2.0*AD*AI3+(2.0*BD-AD)*AI2-BD*AI1)
          HI=-ALEN*(2.0*E*(2.0*AD*AI23+(2.0*BD-3.0*AD)*AI22+(AD-3.0*BD)*AI21
         &          +BD*AI2O)-DRC*(2.0*AI2-3.0*AI1+AIO))
          HJ=-ALEN*(8.0*E*(-AD*AI23+(AD-BD)*AI22+BD*AI21)
         &          -4.0*DRC*(-AI2+AI1))
          HK=-ALEN*(2.0*E*(2.0*AD*AI23+(2.0*BD-AD)*AI22-BD*AI21)
         &          -DRC*(2.0*AI2-AI1))
          KI=(ALEN/4.0)*(2.0*AD*AL3+(2.0*BD-3.0*AD)*AL2+(AD-3.0*BD)*AL1
         &          +BD*ALO-(AD/2.0+(2.0*BD-3.0*AD)/3.0+(AD-3.0*BD)/2.0+BD))
          KJ=ALEN*(-AD*AL3+(AD-BD)*AL2+BD*AL1-(-AD/4.0+(AD-BD)/3.0+BD/2.0))
          KK=(ALEN/4.0)*(2.0*AD*AL3+(2.0*BD-AD)*AL2-BD*AL1
         &          -(AD/2.0+(2.0*BD-AD)/3.0-BD/2.0))
          LI=(ALEN/4.0)*(2.0*E*(2.0*AD*AI3+(2.0*BD-3.0*AD)*AI2
         &          +(AD-3.0*BD)*AI1+BD*AIO)
         &          +DRC*(2.0*AL2-3.0*AL1+ALO-1.0/6.0))
          LJ=ALEN*(2.0*E*(-AD*AI3+(AD-BD)*AI2+BD*AI1)
         &          +DRC*(-AL2+AL1-1.0/6.0))
          LK=(ALEN/4.0)*(2.0*E*(2.0*AD*AI3+(2.0*BD-AD)*AI2-BD*AI1)
         &          +DRC*(2.0*AL2-AL1-1.0/6.0))
```

```
C
C     ==================================================
C  =  HARMONIC SOURCE TERM OF FORM:                   =
C  =    F(X,Y) = C1* XY + C2 * X + C3 * Y + C4        =
C     ==================================================
C
      IF(IB.EQ.1)THEN
C
      AI6=(0.20-B*AI5-C*AI4)/A
      AL4=(ALNC2-2.0*A*AI6-B*AI5)/5.0
C
      G1=C1*AX*AY
      G2=C1*(AY*XI+AX*YI)+C2*AX+C3*AY
      G3=C1*XI*YI+C2*XI+C3*YI+C4
      H1=C1*(AY*DCX+AX*DCY)
      H2=(C1*YI+C2)*DCX+(C1*XI+C3)*DCY
C
      P1=G1*AD
      P2=G1*BD+G2*AD
      P3=G2*BD+G3*AD
      P4=G3*BD
      Q1=H1*A*AD-A*G1*DRC
      Q2=H1*(A*BD+B*AD)+H2*A*AD-(A*G2+B*G1)*DRC
      Q3=H1*(B*BD+C*AD)+H2*(A*BD+B*AD)-(A*G3+B*G2+C*G1)*DRC
      Q4=H1*C*BD+H2*(B*BD+C*AD)-(B*G3+C*G2)*DRC
      Q5=H2*C*BD-C*G3*DRC
      R1=H1*AD
      R2=H1*BD+H2*AD
      R3=H2*BD
C
      B1=(ALEN/16.0)*(-Q1*AL4/2.0+(P1*E-Q2/2.0)*AL3+(P2*E-Q3/2.0)*AL2
     &             +(P3*E-Q4/2.0)*AL1+(P4*E-Q5/2.0)*ALO
     &             -1.5*E*(P1/4.0+P2/3.0+P3/2.0+P4)
     &             +1.25*(Q1/5.0+Q2/4.0+Q3/3.0+Q4/2.0+Q5))
C
      B2=(ALEN/4.0)*(2.0*E*(P1*AI3+P2*AI2+P3*AI1+P4*AIO)
     &             +(G1*DRC-R1)*AL2+(G2*DRC-R2)*AL1+(G3*DRC-R3)*ALO
     &             -DRC*(G1/3.0+G2/2.0+G3)+(R1/3.0+R2/2.0+R3))
C
      END IF
C
C     =======================================================================
C  =  GENERAL BIHARMONIC SOURCE TERM; DEFINEED IN SUBROUTINE   =
C  =  FUNCTION AND EVALUATED BY GAUSSIAN QUADRATURE            =
C     =======================================================================
C
      IF(IB.EQ.2.OR.IB.EQ.10)THEN
C
C     ============================
C  =  BEGIN SURFACE INTEGRATION =
C     ============================
C
      DO 100 I=1,NUMGP
C
        GKI=GP(I)
        XG=AX*GKI+XI
        YG=AY*GKI+YI
        AARG=AD*GKI+BD
        AARG1=A*GKI*GKI+B*GKI+C
        ALNX=DLOG(AARG1)
C
        B1F=(F(XG,YG)/32.0)*(2.0*E*AARG*(ALNX-1.5)
     &          +DRC*AARG1*(ALNX-2.5))
        B1FN=(FN(XG,YG,DCX,DCY)/32.0)*AARG1*(ALNX-2.5)*AARG
        B1L=(FL(XG,YG)/768.0)*(4.0*E*AARG*AARG1*(ALNX-17.0/6.0)
     &          +DRC*AARG1*AARG1*(ALNX-10.0/3.0))
        B1LN=(FLN(XG,YG,DCX,DCY)/768.0)*AARG1*AARG1
     &          *(ALNX-10.0/3.0)*AARG
```

```
C
        B2F=(F(XG,YG)/4.0)*(2.0*E*AARG/AARG1+DRC*(ALNX-1.0))
        B2FN=(FN(XG,YG,DCX,DCY)/4.0)*AARG*(ALNX-1.0)
        B2L=(FL(XG,YG)/32.0)*(2.0*E*AARG*(ALNX-1.5)
     &       +DRC*AARG1*(ALNX-2.5))
        B2LN=(FLN(XG,YG,DCX,DCY)/32.0)*AARG1*(ALNX-2.5)*AARG
C
        B1=B1+(B1F-B1FN+B1L-B1LN)*ALEN*GW(I)
        B2=B2+(B2F-B2FN+B2L-B2LN)*ALEN*GW(I)
C
  100   CONTINUE
C
      END IF
C
      END IF
      RETURN
      END
```

6.9 Subroutine INTEGO

INTEGO is the Overhauser element version of the integration subroutine necessary to evaluate Equations (2.30) - (2.35). As shown in Chapter 3, the Overhauser element is defined between two "regular" nodal points and utilizes nodes
on adjacent elements to maintain derivative continuity, see Figure 3.5. Although
the integration occurs between the two interior "regular" points, the integration
generates four components that partially account for four corresponding locations
in each of the matrices [G], [H], [K], and [L]. The remainder of each component
in the matrices is accounted for from the adjacent elements through shared boundary nodes.

In general, a boundary node may coincide with either node defining the Overhauser integration or any other discrete boundary node. The formulation presented can accommodate each of these three possible cases and any resulting singularities. The same comments on the characteristics associated with INTEGQ in
terms of the code length and its structure are also valid in describing INTEGO.

For a general curved boundary, the required integrations and the evaluation
of the biharmonic form of the source term $f(x,y)$ are obtained by numerical quadrature. As with all previous element types, an arbitrary form of the function
$f(x,y)$ is evaluated by the domain "fanning" technique implemented in subroutine
FAN.

The Overhauser formulation also exhibits the same inherent characteristics over rectilinear segments of the boundary as in the quadratic element formulation. Analytical expressions derived in Chapter 3 are used for the necessary surface integrations over each linear segment. Also, an exact evaluation of the special harmonic form of the source term f(x,y) is possible if the geometry is demarcated completely by linear segments.

A series of function statements labeled NIO, NJO, NKO, NLO, JO, ARGO, ARGO1, and DCO isolating basic parametric characteristics of the Overhauser element formulation are included at the end of subroutine INTEGO. In this case, the first four functions NIO, NJO, NKO, and NLO define the Overhauser shape functions, JO is the Jacobian, ARGO and ARGO1 define the Overhauser version of both frequently encountered forms of the position vector, and DCO defines both the x and y direction cosines.

```
      SUBROUTINE INTEGO(X,Y,XP,YP,N1,N2,N3,N4,GI,GJ,GK,GL,HI,HJ,HK,HL,
     $                 KI,KJ,KK,KL,LI,LJ,LK,LL,IB,B1,B2)
C
C =================================================================
C  =                                                              =
C  =    THIS SUBROUTINE INTEGRATES THE TERMS OF THE ASSEMBLY      =
C  =    MATRICES USING OVERHAUSER ELEMENTS                        =
C  =                                                              =
C  =    CALLING ARGUMENTS -                                       =
C  =                                                              =
C  =       XP,YP = COORDINATES OF SOURCE POINT                    =
C  =       XI,YI,                                                 =
C  =       XJ,YJ,                                                 =
C  =       XK,YK,                                                 =
C  =       XL,YL = COORDINATES OF ENDNODES OF ELEMENT BEING INTEGRATED =
C  =               IN CCW ORDER                                   =
C  =    HI,HJ,HK = MATRIX TERMS OF 'HKERN' CORRESPONDING TO EACH NODE =
C  =       HL                                                     =
C  =    GI,GJ,GK = MATRIX TERMS OF 'GKERN' CORRESPONDING TO EACH NODE =
C  =       GL                                                     =
C  =    KI,KJ,KK = MATRIX TERMS OF 'KKERN' CORRESPONDING TO EACH NODE =
C  =       KL                                                     =
C  =    LI,LJ,LK = MATRIX TERMS OF 'LKERN' CORRESPONDING TO EACH NODE =
C  =       LL                                                     =
C  =                                                              =
C =================================================================
C
      IMPLICIT REAL*8(A-H,O-Z)
      REAL*8 KI,KJ,KK,KL,LI,LJ,LK,LL,NIO,NJO,NKO,NLO,JO
      DIMENSION X(1),Y(1)
      COMMON/PCON/C1,C2,C3,C4
```

```
      COMMON/GQUAD/GP(100),GW(100),NUMGP
      COMMON/GLQUAD/GLP(7),GLW(7)
      COMMON/MODEL/NG1,NTOT,NTOT2,NUMNP,NUMELE,NUMEQ,NUMINT,ITYPE,IPLOT
      DATA PI/3.1415926535897932/
      DATA SMALL,SMALL1/1.OE-10,1.OE-2/
C
      LINEAR=0
      B1=0.0
      B2=0.0
      GI=0.0
      GJ=0.0
      GK=0.0
      GL=0.0
      HI=0.0
      HJ=0.0
      HK=0.0
      HL=0.0
      KI=0.0
      KJ=0.0
      KK=0.0
      KL=0.0
      LI=0.0
      LJ=0.0
      LK=0.0
      LL=0.0
C
      XI=X(N1)
      YI=Y(N1)
      XJ=X(N2)
      YJ=Y(N2)
      XK=X(N3)
      YK=Y(N3)
      XL=X(N4)
      YL=Y(N4)
C
C ===============================
C =  CHECK FOR LINEAR GEOMETRY  =
C ===============================
C
      A1=XI-XJ
      A2=XK-XJ
      A3=XJ-XK
      A4=XL-XK
      B1=YI-YJ
      B2=YK-YJ
      B3=YJ-YK
      B4=YL-YK
      AREA1=DABS(A1*B2-B1*A2)
      AREA2=DABS(A3*B4-B3*A4)
      IF(AREA1.LE.SMALL.AND.AREA2.LE.SMALL)THEN
        ALEN1=A1*A1+B1*B1
        ALEN2=A2*A2+B2*B2
        ALEN3=A4*A4+B4*B4
      IF((ALEN1-ALEN2).LE.SMALL1.AND.(ALEN2-ALEN3).LE.SMALL1)LINEAR=1
      END IF
C
      IF(LINEAR.EQ.0)THEN
C
C =====================================================
C = ANALYSIS FOR OVERHAUSER GEOMETRY; INTEGRATIONS =
C = PREFORMED BY NUMERICAL QUADRATURE              =
C =====================================================
C
      AX=-0.5*XI+1.5*XJ-1.5*XK+0.5*XL
      AY=-0.5*YI+1.5*YJ-1.5*YK+0.5*YL
      BX=XI-2.5*XJ+2.0*XK-0.5*XL
      BY=YI-2.5*YJ+2.0*YK-0.5*YL
      CX=-0.5*XI+0.5*XK
      CY=-0.5*YI+0.5*YK
```

```
      DX=XJ-XP
      DY=YJ-YP
C
      A=AX*AX+AY*AY
      B=BX*BX+BY*BY
      C=CX*CX+CY*CY
      D=DX*DX+DY*DY
      AB=2.O*(AX*BX+AY*BY)
      AC=2.O*(AX*CX+AY*CY)
      AD=2.O*(AX*DX+AY*DY)
      BC=2.O*(BX*CX+BY*CY)
      BD=2.O*(BX*DX+BY*DY)
      CD=2.O*(CX*DX+CY*DY)
C
C =====================================
C =  ANALYSIS FOR (XP,YP) = (XJ,YJ)  =
C =====================================
C
          XDF=DABS(XP-X(N2))
          YDF=DABS(YP-Y(N2))
          IF(XDF.LE.SMALL.AND.YDF.LE.SMALL)THEN
C
          DO 10 I=1,NUMGP
C
            GPI=GP(I)
            GWI=GW(I)
C
            AJ=JO(GPI,A,B,C,AB,AC,BC)
            XX=ARGO(GPI,A,B,C,D,AB,AC,AD,BC,BD,CD)
            XXT=A*GPI**4 + AB*GPI**3 + (AC+B)*GPI*GPI
     $                              + (AD+BC)*GPI + (BD+C)
            ALNX=DLOG(XX)
            ALNXT=DLOG(XXT)
            AXP=ARGO1(GPI,AX,BX,CX,DX)
            AYP=ARGO1(GPI,AY,BY,CY,DY)
            DCX=DCO(GPI,AY,BY,CY)
            DCY=-DCO(GPI,AX,BX,CX)
            DTERM=AXP*DCX+AYP*DCY
C
            C1S=-O.5*ALNXT*AJ*GWI
            C1=-O.5*ALNX*AJ*GWI
            C2=-(DTERM/XX)*GWI
            C3=-O.125*XX*(ALNX-2.O)*AJ*GWI
            C4=-O.25*(ALNX-1.O)*DTERM*GWI
C
            GI=GI+C1*NIO(GPI)
            GJ=GJ+C1S*NJO(GPI)
            GK=GK+C1*NKO(GPI)
            GL=GL+C1*NLO(GPI)
C
            HI=HI+C2*NIO(GPI)
            HK=HK+C2*NKO(GPI)
            HL=HL+C2*NLO(GPI)
C
            KI=KI+C3*NIO(GPI)
            KJ=KJ+C3*NJO(GPI)
            KK=KK+C3*NKO(GPI)
            KL=KL+C3*NLO(GPI)
C
            LI=LI+C4*NIO(GPI)
            LJ=LJ+C4*NJO(GPI)
            LK=LK+C4*NKO(GPI)
            LL=LL+C4*NLO(GPI)
C
          IF(IB.EQ.2.OR.IB.EQ.10)THEN
C
            ADCX=DCX/AJ
            ADCY=DCY/AJ
            XG=NIO(GPI)*XI+NJO(GPI)*XJ+NKO(GPI)*XK+NLO(GPI)*XL
```

```
         YG=NIO(GPI)*YI+NJO(GPI)*YJ+NKO(GPI)*YK+NLO(GPI)*YL
C
         B1F=(-F(XG,YG)/32.0)*XX*(ALNX-2.5)*DTERM
         B1FN=(-FN(XG,YG,ADCX,ADCY)/128.0)*XX*XX*(ALNX-3.0)*AJ
         B1L=(-FL(XG,YG)/768.0)*DTERM*XX*XX*(ALNX-10.0/3.0)
         B1LN=(-FLN(XG,YG,ADCX,ADCY)/4608.0)*(XX**3)
     &                              *(ALNX-11.0/3.0)*AJ
C
         B2F=(-F(XG,YG)/4.0)*(ALNX-1.0)*DTERM
         B2FN=(-FN(XG,YG,ADCX,ADCY)/8.0)*XX*(ALNX-2.0)*AJ
         B2L=(-FL(XG,YG)/32.0)*XX*(ALNX-2.5)*DTERM
         B2LN=(-FLN(XG,YG,ADCX,ADCY)/128.0)*XX*XX*(ALNX-3.0)*AJ
C
         B1=B1+(B1F-B1FN+B1L-B1LN)*GWI
         B2=B2+(B2F-B2FN+B2L-B2LN)*GWI
C
       END IF
C
  10     CONTINUE
C
         DO 15 I=1,7
C
         GLPI=GLP(I)
         GLWI=GLW(I)
C
         AJ=JO(GLPI,A,B,C,AB,AC,BC)
         C1=-AJ*GLWI
C
         GJ=GJ+NJO(GLPI)*C1
C
  15     CONTINUE
C
       RETURN
       END IF
C
C ======================================
C =  ANALYSIS FOR (XP,YP) = (XK,YK)  =
C ======================================
C
         XDF=DABS(XP-X(N3))
         YDF=DABS(YP-Y(N3))
         IF(XDF.LE.SMALL.AND.YDF.LE.SMALL)THEN
C
         DO 20 I=1,NUMGP
C
         GPI=GP(I)
         GWI=GW(I)
         TS=1.0-GPI
C
         AJ=JO(TS,A,B,C,AB,AC,BC)
         XX=ARGO(TS,A,B,C,D,AB,AC,AD,BC,BD,CD)
         XXT=A*GPI**4 - (6.0*A+AB)*GPI**3 + (15.0*A+5.0*AB
     $       +AC+B)*GPI*GPI - (20.0*A+10.0*AB+4.0*(AC+B)+AD+BC)*GPI
     $       +15.0*A+10.0*AB+6.0*(AC+B)+3.0*(AD+BC)+BD+C
         ALNX=DLOG(XX)
         ALNXT=DLOG(XXT)
         AXP=ARGO1(TS,AX,BX,CX,DX)
         AYP=ARGO1(TS,AY,BY,CY,DY)
         DCX=DCO(TS,AY,BY,CY)
         DCY=-DCO(TS,AX,BX,CX)
         DTERM=AXP*DCX+AYP*DCY
C
         C1S=-0.5*ALNXT*AJ*GWI
         C1=-0.5*ALNX*AJ*GWI
         C2=-(DTERM/XX)*GWI
         C3=-0.125*XX*(ALNX-2.0)*AJ*GWI
         C4=-0.25*(ALNX-1.0)*DTERM*GWI
C
         GI=GI+C1*NIO(TS)
```

```
              GJ=GJ+C1*NJO(TS)
              GK=GK+C1S*NKO(TS)
              GL=GL+C1*NLO(TS)
C
              HI=HI+C2*NIO(TS)
              HJ=HJ+C2*NJO(TS)
              HL=HL+C2*NLO(TS)
C
              KI=KI+C3*NIO(TS)
              KJ=KJ+C3*NJO(TS)
              KK=KK+C3*NKO(TS)
              KL=KL+C3*NLO(TS)
C
              LI=LI+C4*NIO(TS)
              LJ=LJ+C4*NJO(TS)
              LK=LK+C4*NKO(TS)
              LL=LL+C4*NLO(TS)
C
          IF(IB.EQ.2.OR.IB.EQ.10)THEN
C
              ADCX=DCX/AJ
              ADCY=DCY/AJ
C
              XGS=NIO(TS)*XI+NJO(TS)*XJ+NKO(TS)*XK+NLO(TS)*XL
              YGS=NIO(TS)*YI+NJO(TS)*YJ+NKO(TS)*YK+NLO(TS)*YL
C
              B1F=(-F(XGS,YGS)/32.0)*XX*(ALNX-2.5)*DTERM
              B1FN=(-FN(XGS,YGS,ADCX,ADCY)/128.0)*XX*XX*(ALNX-3.0)*AJ
              B1L=(-FL(XGS,YGS)/768.0)*DTERM*XX*XX*(ALNX-10.0/3.0)
              B1LN=(-FLN(XGS,YGS,ADCX,ADCY)/4608.0)*(XX**3)
     &                            *(ALNX-11.0/3.0)*AJ
C
              B2F=(-F(XGS,YGS)/4.0)*(ALNX-1.0)*DTERM
              B2FN=(-FN(XGS,YGS,ADCX,ADCY)/8.0)*XX*(ALNX-2.0)*AJ
              B2L=(-FL(XGS,YGS)/32.0)*XX*(ALNX-2.5)*DTERM
              B2LN=(-FLN(XGS,YGS,ADCX,ADCY)/128.0)*XX*XX*(ALNX-3.0)*AJ
C
              B1=B1+(B1F-B1FN+B1L-B1LN)*GWI
              B2=B2+(B2F-B2FN+B2L-B2LN)*GWI
C
          END IF
C
 20           CONTINUE
C
              DO 25 I=1,7
C
              GLPI=GLP(I)
              GLWI=GLW(I)
              TS=1.0-GLPI
C
              AJ=JO(TS,A,B,C,AB,AC,BC)
              C1=-AJ*GLWI
C
              GK=GK+NKO(TS)*C1
C
 25           CONTINUE
C
          RETURN
          END IF
C
          DO 30 I=1,NUMGP
C
              GPI=GP(I)
              GWI=GW(I)
C
              AJ=JO(GPI,A,B,C,AB,AC,BC)
              XX=ARGO(GPI,A,B,C,D,AB,AC,AD,BC,BD,CD)
              ALNX=DLOG(XX)
              AXP=ARGO1(GPI,AX,BX,CX,DX)
```

```
                 AYP=ARGO1(GPI,AY,BY,CY,DY)
                 DCX=DCO(GPI,AY,BY,CY)
                 DCY=-DCO(GPI,AX,BX,CX)
                 DTERM=AXP*DCX+AYP*DCY
C
                 C1=-0.5*ALNX*AJ*GWI
                 C2=-(DTERM/XX)*GWI
                 C3=-0.125*XX*(ALNX-2.0)*AJ*GWI
                 C4=-0.25*(ALNX-1.0)*DTERM*GWI
C
                 GI=GI+C1*NIO(GPI)
                 GJ=GJ+C1*NJO(GPI)
                 GK=GK+C1*NKO(GPI)
                 GL=GL+C1*NLO(GPI)
C
                 HI=HI+C2*NIO(GPI)
                 HJ=HJ+C2*NJO(GPI)
                 HK=HK+C2*NKO(GPI)
                 HL=HL+C2*NLO(GPI)
C
                 KI=KI+C3*NIO(GPI)
                 KJ=KJ+C3*NJO(GPI)
                 KK=KK+C3*NKO(GPI)
                 KL=KL+C3*NLO(GPI)
C
                 LI=LI+C4*NIO(GPI)
                 LJ=LJ+C4*NJO(GPI)
                 LK=LK+C4*NKO(GPI)
                 LL=LL+C4*NLO(GPI)
C
                 IF(IB.EQ.2.OR.IB.EQ.10)THEN
C
                 XG=NIO(GPI)*XI+NJO(GPI)*XJ+NKO(GPI)*XK+NLO(GPI)*XL
                 YG=NIO(GPI)*YI+NJO(GPI)*YJ+NKO(GPI)*YK+NLO(GPI)*YL
                 ADCX=DCX/AJ
                 ADCY=DCY/AJ
C
                 B1F=(-F(XG,YG)/32.0)*XX*(ALNX-2.5)*DTERM
                 B1FN=(-FN(XG,YG,ADCX,ADCY)/128.0)*XX*XX*(ALNX-3.0)*AJ
                 B1L=(-FL(XG,YG)/768.0)*DTERM*XX*XX*(ALNX-10.0/3.0)
                 B1LN=(-FLN(XG,YG,ADCX,ADCY)/4608.0)*(XX**3)
        &                             *(ALNX-11.0/3.0)*AJ
C
                 B2F=(-F(XG,YG)/4.0)*(ALNX-1.0)*DTERM
                 B2FN=(-FN(XG,YG,ADCX,ADCY)/8.0)*XX*(ALNX-2.0)*AJ
                 B2L=(-FL(XG,YG)/32.0)*XX*(ALNX-2.5)*DTERM
                 B2LN=(-FLN(XG,YG,ADCX,ADCY)/128.0)*XX*XX*(ALNX-3.0)*AJ
C
                 B1=B1+(B1F-B1FN+B1L-B1LN)*GWI
                 B2=B2+(B2F-B2FN+B2L-B2LN)*GWI
C
                 END IF
C
 30       CONTINUE
C
          RETURN
C
          ELSE
C
C ===========================================
C = OVERHAUSER ANALYSIS FOR LINEAR GEOMETRY; =
C = EXACT INTEGRATIONS                       =
C ===========================================
C
          AX=XK-XJ
          AY=YK-YJ
          BX=XJ-XP
          BY=YJ-YP
          CX=XK-XP
```

```fortran
      CY=YK-YP
C
      A=AX*AX+AY*AY
      B=2*(AX*BX+AY*BY)
      C=BX*BX+BY*BY
      D=CX*CX+CY*CY
      CCC=4*A*C-B*B
      CC2=A+B+C
C
      ALEN=DSQRT(A)
      CLEN=DSQRT(C)
      DLEN=DSQRT(D)
      DCX=AY/ALEN
      DCY=-AX/ALEN
      E=BX*DCX+BY*DCY
C
      IF(CCO.LE.O.O)THEN
        CC1=1.0
        ANGLE=0.0
      ELSE
        CC1=DSQRT(CCO)
        AARG=(BX*CX+BY*CY)/(CLEN*DLEN)
        IF(AARG.GT.1.0)AARG=1.0
        IF(AARG.LT.-1.0)AARG=-1.0
        ANGLE=ACOS(AARG)
      END IF
C
      IF(C.EQ.O.O)THEN
        ALNC=0.0
      ELSE
        ALNC=DLOG(C)
      END IF
C
      IF(CC2.EQ.O.O)THEN
        ALNC2=0.0
      ELSE
        ALNC2=DLOG(CC2)
      END IF
C
      AIO=2.0*ANGLE/CC1
      AI1=(ALNC2-ALNC-B*AIO)/(2.0*A)
      AI2=(1.0-B*AI1-C*AIO)/A
      AI3=(0.5-B*AI2-C*AI1)/A
      AI4=((1.0/3.0)-B*AI3-C*AI2)/A
      AI5=(0.25-B*AI4-C*AI3)/A
      AI6=(0.20-B*AI5-C*AI4)/A
      AI7=((1.0/6.0)-B*AI6-C*AI5)/A
C
      ALO=ALNC2-2.0*A*AI2-B*AI1
      AL1=(ALNC2-2.0*A*AI3-B*AI2)/2.0
      AL2=(ALNC2-2.0*A*AI4-B*AI3)/3.0
      AL3=(ALNC2-2.0*A*AI5-B*AI4)/4.0
      AL4=(ALNC2-2.0*A*AI6-B*AI5)/5.0
      AL5=(ALNC2-2.0*A*AI7-B*AI6)/6.0
C
      GI=-(ALEN/4.0)*(-AL3+2.0*AL2-AL1)
      GJ=-(ALEN/4.0)*(3.0*AL3-5.0*AL2+2.0*ALO)
      GK=-(ALEN/4.0)*(-3.0*AL3+4.0*AL2+AL1)
      GL=-(ALEN/4.0)*(AL3-AL2)
      HI=(-E*ALEN/2.0)*(-AI3+2.0*AI2-AI1)
      HJ=(-E*ALEN/2.0)*(3.0*AI3-5.0*AI2+2.0*AIO)
      HK=(-E*ALEN/2.0)*(-3.0*AI3+4.0*AI2+AI1)
      HL=(-E*ALEN/2.0)*(AI3-AI2)
      KI=-(ALEN/16.0)*(-A*AL5+(2.0*A-B)*AL4+(2.0*B-A-C)*AL3+(2.0*C-B)
     &          *AL2-C*AL1-2.0*(-A/6.0+(2.0*A-B)/5.0+(2.0*B-A-C)/4.0
     &          +(2.0*C-B)/3.0-C/2.0))
      KJ=-(ALEN/16.0)*(3.0*A*AL5+(3.0*B-5.0*A)*AL4+(3.0*C-5.0*B)*AL3
     &          +(2.0*A-5.0*C)*AL2+2.0*B*AL1+2.0*C*ALO-2.0*(A/2.0
     &          +(3.0*B-5.0*A)/5.0+(3.0*C-5.0*B)/4.0
```

```
     &                  +(2.0*A-5.0*C)/3.0+B+2.0*C))
      KK=-(ALEN/16.0)*(-3.0*A*AL5+(4.0*A-3.0*B)*AL4+(A+4.0*B-3.0*C)
     &              *AL3+(B+4.0*C)*AL2+C*AL1-2.0*(-A/2.0+(4.0*A-3.0*B)
     &               /5.0+(A+4.0*B-3.0*C)/4.0+(B+4.0*C)/3.0+C/2.0))
      KL=-(ALEN/16.0)*(A*AL5+(B-A)*AL4+(C-B)*AL3-C*AL2-2.0*(A/6.0
     &              +(B-A)/5.0+(C-B)/4.0-C/3.0))
      LI=-E*(ALEN/8.0)*(-AL3+2.0*AL2-AL1+1.0/12.0)
      LJ=-E*(ALEN/8.0)*(3.0*AL3-5.0*AL2+2.0*ALO-13.0/12.0)
      LK=-E*(ALEN/8.0)*(-3.0*AL3+4.0*AL2+AL1-13.0/12.0)
      LL=-E*(ALEN/8.0)*(AL3-AL2+1.0/12.0)
C
C     ==============================================
C     =  HARMONIC SOURCE TERM OF FORM:             =
C     =   F(X,Y) = C1* XY + C2 * X + C3 * Y + C4   =
C     ==============================================
C
      IF(IB.EQ.1)THEN
C
      S1=A/3.0+B/2.0+C
      S2=A/4.0+B/3.0+C/2.0
      S3=A/5.0+B/4.0+C/3.0
      P1=A*A
      P2=2*A*B
      P3=2*A*C+B*B
      P4=2*B*C
      P5=C*C
      T1=P1/5.0+P2/4.0+P3/3.0+P4/2.0+P5
      T2=P1/6.0+P2/5.0+P3/4.0+P4/3.0+P5/2.0
C
      G1=C1*AX*AY
      G2=C1*(AY*XJ+AX*YJ)+C2*AX+C3*AY
      G3=C1*XJ*YJ+C2*XJ+C3*YJ+C4
      H1=C1*(AY*DCX+AX*DCY)
      H2=(C1*YJ+C2)*DCX+(C1*XJ+C3)*DCY
C
      B1=(-ALEN/32.0)*(-H1*P1*AL5/4.0+(G1*E*A-(H1*P2+H2*P1)/4.0)*AL4
     &               +(E*(G1*B+G2*A)-(H1*P3+H2*P2)/4.0)*AL3
     &               +(E*(G1*C+G2*B+G3*A)-(H1*P4+H2*P3)/4.0)*AL2
     &               +(E*(G2*C+G3*B)-(H1*P5+H2*P4)/4.0)*AL1
     &               +(E*G3*C-(H2*P5/4.0))*ALO
     &               -(2.5*E)*(G1*S3+G2*S2+G3*S1)
     &               +0.75*(H1*T2+H2*T1))
C
      B2=(-ALEN/4.0)*((-H1*A/2.0)*AL3+(G1*E-(H1*B+H2*A)/2.0)*AL2
     &               +(G2*E-(H1*C+H2*B)/2.0)*AL1
     &               +(G3*E-(H2*C)/2.0)*ALO
     &               -E*(G1/3.0+G2/2.0+G3)+(H1*S2+H2*S1))
C
      END IF
C
C     ==============================================
C     = BEGIN SURFACE INTEGRATION FOR SOURCE TERM =
C     ==============================================
C
      IF(IB.EQ.2.OR.IB.EQ.10)THEN
C
      DO 100 I=1,NUMGP
C
         GKI=GP(I)
         XG=AX*GKI+XJ
         YG=AY*GKI+YJ
         AARG1=A*GKI*GKI+B*GKI+C
         ALNX=DLOG(AARG1)
C
         B1F=(-F(XG,YG)*E/32.0)*AARG1*(ALNX-2.5)
         B1FN=(-FN(XG,YG,DCX,DCY)/128.0)*AARG1*AARG1*(ALNX-3.0)
         B1L=(-FL(XG,YG)/768.0)*E*AARG1*AARG1*(ALNX-10.0/3.0)
         B1LN=(-FLN(XG,YG,DCX,DCY)/4608.0)*(AARG1**3)
     &                            *(ALNX-11.0/3.0)
```

```
C
        B2F=(-F(XG,YG)*E/4.O)*(ALNX-1.O)
        B2FN=(-FN(XG,YG,DCX,DCY)/8.O)*AARG1*(ALNX-2.O)
        B2L=(-FL(XG,YG)*E/32.O)*AARG1*(ALNX-2.5)
        B2LN=(-FLN(XG,YG,DCX,DCY)/128.O)*AARG1*AARG1*(ALNX-3.O)
C
        B1=B1+(B1F-B1FN+B1L-B1LN)*ALEN*GW(I)
        B2=B2+(B2F-B2FN+B2L-B2LN)*ALEN*GW(I)
C
  100   CONTINUE
        END IF
C
        END IF
        RETURN
        END
C
C----------------------------------------------- NIO SHAPE FUNCTION
C
        FUNCTION NIO(T)
        IMPLICIT REAL*8(A-H,O-Z)
        REAL*8 NIO
C
        TS=T*T
        NIO = -0.5*TS*T + TS - 0.5*T
C
        RETURN
        END
C
C----------------------------------------------- NJO SHAPE FUNCTION
C
        FUNCTION NJO(T)
        IMPLICIT REAL*8(A-H,O-Z)
        REAL*8 NJO
C
        TS=T*T
        NJO = 1.5*TS*T - 2.5*TS + 1.0
C
        RETURN
        END
C
C----------------------------------------------- NKO SHAPE FUNCTION
C
        FUNCTION NKO(T)
        IMPLICIT REAL*8(A-H,O-Z)
        REAL *8 NKO
C
        TS= T*T
        NKO = -1.5*TS*T + 2.0*TS + 0.5*T
C
        RETURN
        END
C
C----------------------------------------------- NLO SHAPE FUNCTION
C
        FUNCTION NLO(T)
        IMPLICIT REAL*8(A-H,O-Z)
        REAL *8 NLO
C
        TS= T*T
        NLO = 0.5*TS*T - 0.5*TS
C
        RETURN
        END
C
C----------------------------------------------- JACOBIAN
C
        FUNCTION JO(T,A,B,C,AB,AC,BC)
        IMPLICIT REAL*8(A-H,O-Z)
        REAL*8 JO
```

```
c
         TS=T*T
         JD = DSQRT( 9.0*A*TS*TS + 6.0*AB*TS*T + (3.0*AC+4.0*B)*TS
     $                   + 2.0*BC*T + C )
c
         RETURN
         END
c
c----------------------------------------------- FUNCTION ARGD
c
         FUNCTION ARGD(T,A,B,C,D,AB,AC,AD,BC,BD,CD)
         IMPLICIT REAL*8(A-H,O-Z)
c
         TS=T*T
         TC=TS*T
         ARGD = A*(TC*TC) + AB*TC*TS + (AC+B)*TS*TS + (AD+BC)*TC
     $               + (BD+C)*TS + CD*T + D
c
         RETURN
         END
c
c----------------------------------------------- FUNCTION ARGD1
c
         FUNCTION ARGD1(T,A,B,C,D)
         IMPLICIT REAL*8(A-H,O-Z)
c
         TS=T*T
         ARGD1 = A*TS*T + B*TS + C*T + D
c
         RETURN
         END
c
c---------------------------------------- DIRECTION COSINE FUNCTION
c
         FUNCTION DCO(T,A,B,C)
         IMPLICIT REAL*8(A-H,O-Z)
c
         DCO = 3.0*A*T*T + 2.0*B*T + C
c
         RETURN
         END
```

6.10 Subroutine INTGOD

INTGOD is designed to duplicate the computation of the coordinate derivatives at internal points performed in subroutines INTGQD and INTEGD, except that
INTGOD uses an Overhauser element formulation. The subroutine also contains all
the implicit characteristics described in INTEGQ for general and rectilinear geometries.

```
      SUBROUTINE INTGOD(X,Y,XP,YP,N1,N2,N3,N4,GI,GJ,GK,GL,HI,HJ,HK,HL,
     $                  KI,KJ,KK,KL,LI,LJ,LK,LL,IB,B1,B2,ID)
C
C ======================================================================
C = THIS SUBROUTINE INTEGRATES THE TERMS INVOLVED WITH THE             =
C = X AND Y-DERIVATIVES AT INTERNAL POINTS USING OVERHAUSER ELEMENTS   =
C ======================================================================
C
      IMPLICIT REAL*8(A-H,O-Z)
      REAL*8 KI,KJ,KK,KL,LI,LJ,LK,LL,NIO,NJO,NKO,NLO,JO
      DIMENSION X(1),Y(1)
      COMMON/PCON/C1,C2,C3,C4
      COMMON/GQUAD/GP(100),GW(100),NUMGP
      COMMON/MODEL/NG1,NTOT,NTOT2,NUMNP,NUMELE,NUMEQ,NUMINT,ITYPE,IPLOT
      DATA PI/3.1415926535897932/
      DATA SMALL,SMALL1/1.0E-10,1.0E-2/
C
      LINEAR=0
      B1=0.0
      B2=0.0
      GI=0.0
      GJ=0.0
      GK=0.0
      GL=0.0
      HI=0.0
      HJ=0.0
      HK=0.0
      HL=0.0
      KI=0.0
      KJ=0.0
      KK=0.0
      KL=0.0
      LI=0.0
      LJ=0.0
      LK=0.0
      LL=0.0
C
      XI=X(N1)
      YI=Y(N1)
      XJ=X(N2)
      YJ=Y(N2)
      XK=X(N3)
      YK=Y(N3)
      XL=X(N4)
      YL=Y(N4)
C
C ===============================
C = CHECK FOR LINEAR GEOMETRY =
C ===============================
C
      A1=XI-XJ
      A2=XK-XJ
      A3=XJ-XK
      A4=XL-XK
      B1=YI-YJ
      B2=YK-YJ
      B3=YJ-YK
      B4=YL-YK
      AREA1=DABS(A1*B2-B1*A2)
      AREA2=DABS(A3*B4-B3*A4)
      IF(AREA1.LE.SMALL.AND.AREA2.LE.SMALL)THEN
        ALEN1=A1*A1+B1*B1
        ALEN2=A2*A2+B2*B2
        ALEN3=A4*A4+B4*B4
      IF((ALEN1-ALEN2).LE.SMALL1.AND.(ALEN2-ALEN3).LE.SMALL1)LINEAR=1
      END IF
C
      IF(LINEAR.EQ.0)THEN
C
```

```
C ======================================================
C = ANALYSIS FOR OVERHAUSER GEOMETRY; INTEGRATIONS =
C = PREFORMED BY NUMERICAL QUADRATURE                 =
C ======================================================
C
        AX=-0.5*XI+1.5*XJ-1.5*XK+0.5*XL
        AY=-0.5*YI+1.5*YJ-1.5*YK+0.5*YL
        BX=XI-2.5*XJ+2.0*XK-0.5*XL
        BY=YI-2.5*YJ+2.0*YK-0.5*YL
        CX=-0.5*XI+0.5*XK
        CY=-0.5*YI+0.5*YK
        DX=XJ-XP
        DY=YJ-YP
C
        A=AX*AX+AY*AY
        B=BX*BX+BY*BY
        C=CX*CX+CY*CY
        D=DX*DX+DY*DY
        AB=2.0*(AX*BX+AY*BY)
        AC=2.0*(AX*CX+AY*CY)
        AD1=2.0*(AX*DX+AY*DY)
        BC=2.0*(BX*CX+BY*CY)
        BD=2.0*(BX*DX+BY*DY)
        CD=2.0*(CX*DX+CY*DY)
C
C =============================
C = BEGIN SURFACE QUADRATURE =
C =============================
C
          DO 40 I=1,NUMGP
C
            GPI=GP(I)
            GWI=GW(I)
C
            AJ=JO(GPI,A,B,C,AB,AC,BC)
            XX=ARGO(GPI,A,B,C,D,AB,AC,AD1,BC,BD,CD)
            ALNX=DLOG(XX)
            AXP=ARGO1(GPI,AX,BX,CX,DX)
            AYP=ARGO1(GPI,AY,BY,CY,DY)
            DCX=DCO(GPI,AY,BY,CY)/AJ
            DCY=-DCO(GPI,AX,BX,CX)/AJ
            DTERM=AXP*DCX+AYP*DCY
C
C =========================================
C = DECLARE X OR Y DERIVATIVE BASED ON ID =
C =========================================
C
        IF(ID.EQ.1)THEN
          AD=AXP
          DRC=DCX
        ELSE
          AD=AYP
          DRC=DCY
        END IF
C
            C1=(AD/XX)*AJ*GWI
            C2=(DRC/XX-2.0*AD*DTERM/(XX*XX))*GWI*AJ
            C3=0.25*AD*(ALNX-1.0)*AJ*GWI
            C4=0.25*(2.0*AD*DTERM/XX+DRC*(ALNX-1.0))*GWI*AJ
C
            GI=GI+C1*NIO(GPI)
            GJ=GJ+C1*NJO(GPI)
            GK=GK+C1*NKO(GPI)
            GL=GL+C1*NLO(GPI)
C
            HI=HI+C2*NIO(GPI)
            HJ=HJ+C2*NJO(GPI)
            HK=HK+C2*NKO(GPI)
            HL=HL+C2*NLO(GPI)
```

211

```
C
              KI=KI+C3*NIO(GPI)
              KJ=KJ+C3*NJO(GPI)
              KK=KK+C3*NKO(GPI)
              KL=KL+C3*NLO(GPI)
C
              LI=LI+C4*NIO(GPI)
              LJ=LJ+C4*NJO(GPI)
              LK=LK+C4*NKO(GPI)
              LL=LL+C4*NLO(GPI)
C
          IF(IB.EQ.2.OR.IB.EQ.10)THEN
C
              XG=NIO(GPI)*XI+NJO(GPI)*XJ+NKO(GPI)*XK+NLO(GPI)*XL
              YG=NIO(GPI)*YI+NJO(GPI)*YJ+NKO(GPI)*YK+NLO(GPI)*YL
C
              B1F=(F(XG,YG)/32.0)*(XX*(ALNX-2.5)*DRC+2.0*AD*DTERM
     &                           *(ALNX-1.5))
              B1FN=(FN(XG,YG,DCX,DCY)/32.0)*AD*XX*(ALNX-2.5)
              B1L=(FL(XG,YG)/768.0)*(4.0*AD*DTERM*XX*(ALNX-17.0/6.0)
     &                           +DRC*XX*XX*(ALNX-10.0/3.0))
              B1LN=(FLN(XG,YG,DCX,DCY)/768.0)*XX*XX*AD*(ALNX-10.0/3.0)
C
              B2F=(F(XG,YG)/4.0)*(DRC*(ALNX-1.0)+2.0*AD*DTERM/XX)
              B2FN=(FN(XG,YG,DCX,DCY)/4.0)*AD*(ALNX-1.0)
              B2L=(FL(XG,YG)/32.0)*(XX*(ALNX-2.5)*DRC+2.0*AD*DTERM
     &                           *(ALNX-1.5))
              B2LN=(FLN(XG,YG,DCX,DCY)/32.0)*AD*XX*(ALNX-2.5)
C
              B1=B1+(B1F-B1FN+B1L-B1LN)*GWI*AJ
              B2=B2+(B2F-B2FN+B2L-B2LN)*GWI*AJ
C
          END IF
C
 40       CONTINUE
          RETURN
C
      ELSE
C
C ===========================================
C = OVERHAUSER ANALYSIS FOR LINEAR GEOMETRY; =
C = EXACT INTEGRATIONS                       =
C ===========================================
C
      AX=XK-XJ
      AY=YK-YJ
      BX=XJ-XP
      BY=YJ-YP
      CX=XK-XP
      CY=YK-YP
C
      A=AX*AX+AY*AY
      B=2*(AX*BX+AY*BY)
      C=BX*BX+BY*BY
      D=CX*CX+CY*CY
      CCO=4*A*C-B*B
      CC2=A+B+C
C
      ALEN=DSQRT(A)
      CLEN=DSQRT(C)
      DLEN=DSQRT(D)
      DCX=AY/ALEN
      DCY=-AX/ALEN
      E=BX*DCX+BY*DCY
C
C ===========================================
C = DECLARE X OR Y DERIVATIVE BASED ON ID   =
C ===========================================
C
```

```
        IF(ID.EQ.1)THEN
          AD=AX
          BD=BX
          DRC=DCX
        ELSE
          AD=AY
          BD=BY
          DRC=DCY
        END IF
C
        IF(CCO.LE.O.O)THEN
          CC1=1.O
          ANGLE=O.O
        ELSE
          CC1=DSQRT(CCO)
          AARG=(BX*CX+BY*CY)/(CLEN*DLEN)
          IF(AARG.GT.1.O)AARG=1.O
          IF(AARG.LT.-1.O)AARG=-1.O
          ANGLE=ACOS(AARG)
        END IF
C
        IF(C.EQ.O.O)THEN
          ALNC=O.O
        ELSE
          ALNC=DLOG(C)
        END IF
C
        IF(CC2.EQ.O.O)THEN
          ALNC2=O.O
        ELSE
          ALNC2=DLOG(CC2)
        END IF
C
        AIO=2.O*ANGLE/CC1
        AI1=(ALNC2-ALNC-B*AIO)/(2.O*A)
        AI2=(1.O-B*AI1-C*AIO)/A
        AI3=(O.5-B*AI2-C*AI1)/A
        AI4=((1.O/3.O)-B*AI3-C*AI2)/A
        AI5=(O.25-B*AI4-C*AI3)/A
        AI6=(O.2O-B*AI5-C*AI4)/A
C
        ALO=ALNC2-2.O*A*AI2-B*AI1
        AL1=(ALNC2-2.O*A*AI3-B*AI2)/2.O
        AL2=(ALNC2-2.O*A*AI4-B*AI3)/3.O
        AL3=(ALNC2-2.O*A*AI5-B*AI4)/4.O
        AL4=(ALNC2-2.O*A*AI6-B*AI5)/5.O
C
        IF(CCO.EQ.O.O)THEN
          BA=B/(2.O*A)
          BA3=BA**3
          BD3=(1.O+BA)**3
          AS=A*A
          AI2O=-(1.O/(3.O*BA*BD3)-1.O/(3.O*BA*BA3))/AS
          AI21=-1.O/(2.O*BD3*AS)+(BA/2.O)*AI2O
C         AI22=-1.O/(BD3*AS)+2.O*BA*AI21
        ELSE
          AI2O=((2.O*A+B)/CC2-B/C+2.O*A*AIO)/CCO
          AI21=((2.O*C+B)/CC2-2.O*B*AIO)/(-CCO)
C         AI22=((B*B-2.O*A*C+B*C)/(A*CC2)-B/A+2.O*C*AIO)/CCO
        END IF
          AI22=-1.O/(A*CC2)+(C/A)*AI2O
          AI23=AI1/A-(C/A)*AI21-(B/A)*AI22
          AI24=1.O/(A*CC2)-(2.O*B/A)*AI23-(3.O*C/A)*AI22
C
        GI=(ALEN*O.5)*(-AD*AI4+(2.O*AD-BD)*AI3+(2.O*BD-AD)*AI2-BD*AI1)
        GJ=(ALEN*O.5)*(3.O*AD*AI4+(3.O*BD-5.O*AD)*AI3-5.O*BD*AI2
     &               +2.O*AD*AI1+2.O*BD*AIO)
        GK=(ALEN*O.5)*(-3.O*AD*AI4+(4.O*AD-3.O*BD)*AI3
     &               +(AD+4.O*BD)*AI2+BD*AI1)
```

```
      GL=(ALEN*0.5)*(AD*AI4+(BD-AD)*AI3-BD*AI2)
C
      HI=(-ALEN*0.5)*(2.0*E*(-AD*AI24+(2.0*AD-BD)*AI23+(2.0*BD-AD)*AI22
     &               -BD*AI21)-DRC*(-AI3+2.0*AI2-AI1))
      HJ=(-ALEN*0.5)*(2.0*E*(3.0*AD*AI24+(3.0*BD-5.0*AD)*AI23
     &               -5.0*BD*AI22+2.0*AD*AI21+2.0*BD*AI20)
     &               -DRC*(3.0*AI3-5.0*AI2+2.0*AIO))
      HK=(-ALEN*0.5)*(2.0*E*(-3.0*AD*AI24+(4.0*AD-3.0*BD)*AI23
     &               +(AD+4.0*BD)*AI22+BD*AI21)
     &               -DRC*(-3.0*AI3+4.0*AI2+AI1))
      HL=(-ALEN*0.5)*(2.0*E*(AD*AI24+(BD-AD)*AI23-BD*AI22)
     &               -DRC*(AI3-AI2))
C
      KI=(ALEN/8.0)*(-AD*AL4+(2.0*AD-BD)*AL3+(2.0*BD-AD)*AL2-BD*AL1
     &               -(-AD/5.0+(2.0*AD-BD)/4.0+(2.0*BD-AD)/3.0-BD/2.0))
      KJ=(ALEN/8.0)*(3.0*AD*AL4+(3.0*BD-5.0*AD)*AL3-5.0*BD*AL2
     &               +2.0*AD*AL1+2.0*BD*ALO-(0.6*AD+(3.0*BD-5.0*AD)/4.0
     &               -5.0*BD/3.0+AD+2.0*BD))
      KK=(ALEN/8.0)*(-3.0*AD*AL4+(4.0*AD-3.0*BD)*AL3+(AD+4.0*BD)*AL2
     &               +BD*AL1-(-0.6*AD+(4.0*AD-3.0*BD)/4.0
     &               +(AD+4.0*BD)/3.0+BD/2.0))
      KL=(ALEN/8.0)*(AD*AL4+(BD-AD)*AL3-BD*AL2
     &               -(AD/5.0+(BD-AD)/4.0-BD/3.0))
C
      LI=(ALEN/8.0)*(2.0*E*(-AD*AI4+(2.0*AD-BD)*AI3+(2.0*BD-AD)*AI2
     &               -BD*AI1)+DRC*(-AL3+2.0*AL2-AL1+1.0/12.0))
      LJ=(ALEN/8.0)*(2.0*E*(3.0*AD*AI4+(3.0*BD-5.0*AD)*AI3-5.0*BD*AI2
     &               +2.0*AD*AI1+2.0*BD*AIO)+DRC*(3.0*AL3-5.0*AL2
     &               +2.0*ALO-13.0/12.0))
      LK=(ALEN/8.0)*(2.0*E*(-3.0*AD*AI4+(4.0*AD-3.0*BD)*AI3
     &               +(AD+4.0*BD)*AI2+BD*AI1)+DRC*(-3.0*AL3+4.0*AL2
     &               +AL1-13.0/12.0))
      LL=(ALEN/8.0)*(2.0*E*(AD*AI4+(BD-AD)*AI3-BD*AI2)
     &               +DRC*(AL3-AL2+1.0/12.0))
C
C  ==========================================
C  =  HARMONIC SOURCE TERM OF FORM:         =
C  =   F(X,Y) = C1* XY + C2 * X + C3 * Y + C4  =
C  ==========================================
C
      IF(IB.EQ.1)THEN
C
      G1=C1*AX*AY
      G2=C1*(AY*XJ+AX*YJ)+C2*AX+C3*AY
      G3=C1*XJ*YJ+C2*XJ+C3*YJ+C4
      H1=C1*(AY*DCX+AX*DCY)
      H2=(C1*YJ+C2)*DCX+(C1*XJ+C3)*DCY
C
      P1=G1*AD
      P2=G1*BD+G2*AD
      P3=G2*BD+G3*AD
      P4=G3*BD
      Q1=H1*A*AD-A*G1*DRC
      Q2=H1*(A*BD+B*AD)+H2*A*AD-(A*G2+B*G1)*DRC
      Q3=H1*(B*BD+C*AD)+H2*(A*BD+B*AD)-(A*G3+B*G2+C*G1)*DRC
      Q4=H1*C*BD+H2*(B*BD+C*AD)-(B*G3+C*G2)*DRC
      Q5=H2*C*BD-C*G3*DRC
      R1=H1*AD
      R2=H1*BD+H2*AD
      R3=H2*BD
C
      B1=(ALEN/16.0)*(-Q1*AL4/2.0+(P1*E-Q2/2.0)*AL3+(P2*E-Q3/2.0)*AL2
     &               +(P3*E-Q4/2.0)*AL1+(P4*E-Q5/2.0)*ALO
     &               -1.5*E*(P1/4.0+P2/3.0+P3/2.0+P4)
     &               +1.25*(Q1/5.0+Q2/4.0+Q3/3.0+Q4/2.0+Q5))
C
      B2=(ALEN/4.0)*(2.0*E*(P1*AI3+P2*AI2+P3*AI1+P4*AIO)
     &               +(G1*DRC-R1)*AL2+(G2*DRC-R2)*AL1+(G3*DRC-R3)*ALO
     &               -DRC*(G1/3.0+G2/2.0+G3)+(R1/3.0+R2/2.0+R3))
```

```
C
          END IF
C
C     ================================================================
C     =  GENERAL BIHARMONIC SOURCE TERM; DEFINED IN SUBROUTINE    =
C     =  FUNCTION AND EVALUATED BY GAUSSIAN QUADRATURE             =
C     ================================================================
C
          IF(IB.EQ.2.OR.IB.EQ.10)THEN
C
C     =============================
C     = BEGIN SURFACE INTEGRATION =
C     =============================
C
          DO 100 I=1,NUMGP
C
            GKI=GP(I)
            XG=AX*GKI+XJ
            YG=AY*GKI+YJ
            AARG=AD*GKI+BD
            AARG1=A*GKI*GKI+B*GKI+C
            ALNX=DLOG(AARG1)
C
            B1F=(F(XG,YG)/32.0)*(2.0*E*AARG*(ALNX-1.5)
     &          +DRC*AARG1*(ALNX-2.5))
            B1FN=(FN(XG,YG,DCX,DCY)/32.0)*AARG1*(ALNX-2.5)*AARG
            B1L=(FL(XG,YG)/768.0)*(4.0*AARG*E*AARG1*(ALNX-17.0/6.0)
     &          +DRC*AARG1*AARG1*(ALNX-10.0/3.0))
            B1LN=(FLN(XG,YG,DCX,DCY)/768.0)*AARG1*AARG1*AARG
     &          *(ALNX-10.0/3.0)
C
            B2F=(F(XG,YG)/4.0)*(2*E*AARG/AARG1+DRC*(ALNX-1.0))
            B2FN=(FN(XG,YG,DCX,DCY)/4.0)*AARG*(ALNX-1.0)
            B2L=(FL(XG,YG)/32.0)*(2.0*E*AARG*(ALNX-1.5)
     &          +DRC*AARG1*(ALNX-2.5))
            B2LN=(FLN(XG,YG,DCX,DCY)/32.0)*AARG1*(ALNX-2.5)*AARG
C
            B1=B1+(B1F-B1FN+B1L-B1LN)*ALEN*GW(I)
            B2=B2+(B2F-B2FN+B2L-B2LN)*ALEN*GW(I)
C
  100     CONTINUE
C
          END IF
C
          RETURN
          END IF
          RETURN
          END
```

6.11 Subroutine FAN

FAN is formulated to evaluate an arbitrary form of the source function f(x,y) over a general problem domain. As described in detail in Chapter 4, the technique combines the convenience of higher order triangular quadrature with the inherent advantages of non-discretization of the domain. The algorithm divides the domain into a series of triangular regions formed implicitly by three vertices, two of which are consecutive boundary nodes, and the third is the general field point under consideration. The subroutine can distinguish between the four different element types and adjust the distribution of quadrature points accordingly.

The general subroutine is designed to compute both the domain integrations involving the source term f(x,y) defined in Equations (2.11) and (2.12) for the formation of the vectors B1 and B2 , as well as their x and y coordinate derivatives given in Equations (2.15), (2.16), (2.21), and (2.22) for internal point calculations.

Included in the subroutine FAN are BLOCK DATA TRIGP1 and BLOCK DATA TRIGP2 which store respectively the data for a 7-point and a 13-point Gaussian quadrature for a general triangular region. The Gauss points and weights are converted to a parameterized triangle from the symmetric triangular coordinates in which the points where originally defined (Cowper, 1973).

```
      SUBROUTINE FAN(XP,YP,X,Y,NELE,BT1,BT2,B1X,B1Y,B2X,B2Y)
C
C ================================================================
C =                                                              =
C =   THIS SUBROUTINE INTEGRATES AN ARBITRARY SOURCE TERM FUNCTION =
C =   USING AN IMPROVED DOMAIN "FANNING" TECHNIQUE.  THE SOURCE TERM =
C =   SHOULD BE DEFINED IN THE FUNCTION STATEMENT F(X,Y).        =
C =                                                              =
C ================================================================
C
      IMPLICIT REAL*8(A-H,O-Z)
      DIMENSION XCENT(1),YCENT(1),X(1),Y(1),XT(9),YT(9),
     &          LT(21),NELE(NTOT,4)
      COMMON/TGQ1/ CHI1(7), ETA1(7), WTT1(7)
      COMMON/TGQ2/ CHI2(13), ETA2(13), WTT2(13)
      COMMON/MODEL/NG1,NTOT,NTOT2,NUMNP,NUMELE,NUMEQ,NUMINT,ITYPE,IPLOT
      DATA LT /1,2,3,2,4,3,3,4,8,3,8,5,5,4,6,5,6,9,5,9,7/
      DATA SMALL/1.0E-10/
C
          BT1=0.0
          BT2=0.0
          B1X=0.0
          B2X=0.0
          B1Y=0.0
          B2Y=0.0
          AREA1=0.0
          XT(1)=XP
          YT(1)=YP
C
      DO 200 I=1,NUMELE
C
          IJUMP=0
          N1=NELE(I,1)
```

```
              N2=NELE(I,2)
              N3=NELE(I,3)
              N4=NELE(I,4)
C
              IF(N4.NE.O)THEN
                IJUMP=1
                N1=N2
                N2=N3
              END IF
C
 1000         XT(6)=X(N1)
              XT(7)=X(N2)
              YT(6)=Y(N1)
              YT(7)=Y(N2)
C
              A1C=XT(6)-XP
              A2C=XT(7)-XP
              B1C=YT(6)-YP
              B2C=YT(7)-YP
              AREAC=0.5*(A1C*B2C-A2C*B1C)
              IF(AREAC.LT.SMALL)GOTO 200
C
C     ========================================================
C     =  DIVIDE EACH ELEMENT INTO SEVEN SMALLER TRIANGLES    =
C     ========================================================
C
              DXI=X(N1)-XP
              DYI=Y(N1)-YP
              DXJ=X(N2)-XP
              DYJ=Y(N2)-YP
C
              XT(2)=XP+DXI/8.0
              XT(3)=XP+DXJ/8.0
              YT(2)=YP+DYI/8.0
              YT(3)=YP+DYJ/8.0
              XT(4)=XP+DXI/2.0
              XT(5)=XP+DXJ/2.0
              YT(4)=YP+DYI/2.0
              YT(5)=YP+DYJ/2.0
              XT(8)=(XT(5)+XT(4))/2.0
              XT(9)=(XT(7)+XT(6))/2.0
              YT(8)=(YT(5)+YT(4))/2.0
              YT(9)=(YT(7)+YT(6))/2.0
C
              DO 250 NF=1,21,3
C
              XK=XT(LT(NF))
              YK=YT(LT(NF))
              XI=XT(LT(NF+1))
              YI=YT(LT(NF+1))
              XJ=XT(LT(NF+2))
              YJ=YT(LT(NF+2))
              SUM1=0.0
              SUM2=0.0
              SUM1X=0.0
              SUM2X=0.0
              SUM1Y=0.0
              SUM2Y=0.0
C
C     ==============================
C     =   BEGIN AREA INTEGRATION    =
C     ==============================
C
              NUMTQP=7
              IF(NF.EQ.1)NUMTQP=13
C
              DO 300 NGP=1,NUMTQP
C
              IF(NUMTQP.EQ.7)THEN
```

```
          CHIG=CHI1(NGP)
          ETAG=ETA1(NGP)
          WT=WTT1(NGP)
        ELSE
          CHIG=CHI2(NGP)
          ETAG=ETA2(NGP)
          WT=WTT2(NGP)
        END IF
C
        XG=(1.O-CHIG-ETAG)*XI+CHIG*XJ+ETAG*XK
        YG=(1.O-CHIG-ETAG)*YI+CHIG*YJ+ETAG*YK
        DX=XG-XP
        DY=YG-YP
        ARG=DX*DX+DY*DY
        ALNX=DLOG(ARG)
C
        SUM1=SUM1-F(XG,YG)*WT*ARG*(ALNX-2.0)*0.125
        SUM2=SUM2-F(XG,YG)*WT*ALNX*0.5
        SUM1X=SUM1X+F(XG,YG)*WT*DX*(ALNX-1.0)*0.25
        SUM2X=SUM2X+F(XG,YG)*WT*DX/ARG
        SUM1Y=SUM1Y+F(XG,YG)*WT*DY*(ALNX-1.0)*0.25
        SUM2Y=SUM2Y+F(XG,YG)*WT*DY/ARG
C
 300      CONTINUE
C
C   ======================================
C   =  CALCULATE THE AREA OF THE TRIANGLE  =
C   ======================================
C
        A1=XI-XK
        A2=XJ-XK
        B1=YI-YK
        B2=YJ-YK
        AREA=0.5*(A1*B2-A2*B1)
        AREA1=AREA1+AREA
        BT1=BT1+SUM1*AREA
        BT2=BT2+SUM2*AREA
        B1X=B1X+SUM1X*AREA
        B2X=B2X+SUM2X*AREA
        B1Y=B1Y+SUM1Y*AREA
        B2Y=B2Y+SUM2Y*AREA
C
 250      CONTINUE
C
        IF(IJUMP.EQ.1)GOTO 200
          IF(N3.NE.O.AND.N4.EQ.O)THEN
            N1=N2
            N2=N3
            IJUMP=1
            GOTO 1000
          END IF
C
 200      CONTINUE
C
        RETURN
        END
C
C   ================================================================
C   =   BLOCK DATA FOR GAUSSIAN QUADRATURE POINTS                  =
C   =   FOR A TWO DIMENSIONAL TRIANGULAR AREA   (7 PTS.)           =
C   ================================================================
C
      BLOCK DATA TRIGP1
C
      IMPLICIT REAL*8(A-H,O-Z)
      COMMON/TGQ1/ CHI1(7), ETA1(7), WTT1(7)
C
      DATA CHI1/0.333333333333333,0.797426985353087,
     &          0.797426985353087,0.237932366472434,
```

```
     &              0.237932366472434,0.025355134559132,
     &              0.025355134559132/
          DATA ETA1/0.333333333333333,0.237932366472434,
     &              0.025355134559132,0.736712498968435,
     &              0.025355134559132,0.736712498968435,
     &              0.237932366472434/
          DATA WTT1/0.375000000000000,0.104166666666667,
     &              0.104166666666667,0.104166666666667,
     &              0.104166666666667,0.104166666666667,
     &              0.104166666666667/
C
          END
C
C     ===========================================================
C     =    BLOCK DATA FOR GAUSSIAN QUADRATURE POINTS           =
C     =    FOR A TWO DIMENSIONAL TRIANGULAR AREA   (13 PTS.)   =
C     ===========================================================
C
          BLOCK DATA TRIGP2
C
          IMPLICIT REAL*8(A-H,O-Z)
          COMMON/TGQ2/CHI2(13),ETA2(13),WTT2(13)
C
          DATA CHI2/0.333333333333333,0.479308067841923,
     &              0.260345966079038,0.260345966079038,
     &              0.869739794195568,0.065130102902216,
     &              0.065130102902216,0.638444188569809,
     &              0.638444188569809,0.312865496004875,
     &              0.312865496004875,0.048690315425316,
     &              0.048690315425316/
          DATA ETA2/0.333333333333333,0.260345966079038,
     &              0.479308067841923,0.260345966079038,
     &              0.065130102902216,0.869739794195568,
     &              0.065130102902216,0.312865496004875,
     &              0.048690315425316,0.638444188569809,
     &              0.048690315425316,0.638444188569809,
     &              0.312865496004875/
          DATA WTT2/-0.149570044467670,0.175615257433204,
     &              0.175615257433204,0.175615257433204,
     &              0.053347235608839,0.053347235608839,
     &              0.053347235608839,0.077113760890257,
     &              0.077113760890257,0.077113760890257,
     &              0.077113760890257,0.077113760890257,
     &              0.077113760890257/
C
          END
```

6.12   Subroutines REDSOL, SOLONL, REDUCE, FWDSLV, and BCLSLV

These subroutines combine to comprise a routine capable of performing a full-pivoting Gaussian elimination solution of a system of simultaneous equations. The procedure always attempts to maintain a well-condition state in the assembly matrix at all steps of the solution process. This set of programs is used in here with the permission and courtesy of Professor A. J. McPhate of the Department of Mechanical Engineering at Louisiana State University, Baton Rouge, U.S.A.

```
      SUBROUTINE REDSOL( A,NA,X,Y,N,NRNK )
C
      REAL*8 A(NA,N),X(N),Y(N)
      LOGICAL NRGTO
      NRGTO = NRNK.GT.OO
      CALL REDUCE( A,NA,A(1,N+1),N,A(N-1,N+1),N,NRNK,NP )
      IF( NRNK.EQ.N )                     GOTO  2000
      IF( .NOT.NRGTO )                    RETURN
      DO   1500 I=NRNK+1,N
      A(I,I) = SIGN( 1.0D15,A(I,I) )
      IF( I.EQ.N )                        GOTO  1500
      DO   1000 J=I+1,N
 1000 A(J,I) = 1.0
 1500 CONTINUE
 2000 CALL SOLONL( A,NA,X,Y,N )
                                          RETURN
      END
C
C-----------------------------------------------------
C
      SUBROUTINE SOLONL( A,NA,X,Y,N )
      REAL*8 A(NA,1),X(N),Y(N)
      CALL FWDSLV( A,NA,A(1,N+1),N,N,X,Y )
      DO   3000  I=1,N
 3000 Y(I) = X(I)
      CALL BCKSLV( A,NA,A(N-1,N+1),N,N,X,Y )
                                          RETURN
      END
C
C     A PACKAGE TO REDUCE A MATRIX TO RANK UPPER TRIANGULAR,
C     FORWARD SOLVE A VECTOR TO DETERMINE CONSISTENCY,
C     AND BACK SOLVE TO OBTAIN A SOLUTION.
C
      SUBROUTINE REDUCE( A,NA,ROW,M,COL,N,RANK,NP )
C
      INTEGER NA,M,ROW(M),N,COL(N),RANK,NP,I,J,R,IMX,JMX
      REAL*8 A(NA,N),P,AMX,TEST
C
      DO   1000  I=1,M
 1000 ROW(I) = I
      DO   1100  I=1,N
 1100 COL(I) = I
C
      NP = OO
      RANK = MINO( M,N )
      DO   2000 R=1,RANK
C
      IMX = R
      JMX = R
      AMX = ABS( A(R,R) )
      DO   1200  I=R,M
      DO   1200  J=R,N
      IF( AMX.GE.ABS( A(I,J) ) )        GOTO  1200
      AMX = ABS( A(I,J) )
      IMX =I
      JMX = J
 1200 CONTINUE
C
      IF( R.EQ.1 ) TEST = 1.OE-15*AMX
      IF( AMX.LE.TEST )                 GOTO  3000
C
      IF( IMX.EQ.R )                    GOTO  1400
      DO   1300 J=1,N
      P = A(IMX,J)
      A(IMX,J) = A(R,J)
 1300 A(R,J) = P
      K = ROW(IMX)
      ROW(IMX) = ROW(R)
      ROW(R) = K
```

```
      NP = NP+1
C
 1400 IF( JMX.EQ.R )                        GOTO  1600
      DO   1500 I=1,M
      P = A(I,JMX)
      A(I,JMX) = A(I,R)
 1500 A(I,R) = P
      K = COL(JMX)
      COL(JMX) = COL(R)
      COL(R) = K
      NP = NP+1
C
 1600 AMX = 1.0/A(R,R)
      A(R,R) = AMX
      IF( R.EQ.M )                          GOTO  2000
      DO   1900 I=R+1,M
      P = A(I,R)*AMX
      A(I,R) = P
      IF( R.EQ.N )                          GOTO  1900
      DO   1800 J=R+1,N
 1800 A(I,J) = A(I,J) - P*A(R,J)
 1900 CONTINUE
 2000 CONTINUE
C
                                            GOTO  4000
C
 3000 RANK = R-1
C
 4000                                       RETURN
C
      END
C
C
C
      SUBROUTINE FWDSLV( A,NA,ROW,M,RANK,Y,B1 )
C
      INTEGER NA,M,ROW(M),RANK,I,J
      REAL*8 A(NA,RANK), Y(M),B1(M), P
C
      Y(1) = B1(ROW(1))
      DO   1000 I=2,M
      P = B1(ROW(I))
      K = JMINO( I-1,RANK )
      DO   900 J=1,K
  900 P = P - A(I,J)*Y(J)
 1000 Y(I) = P
C
                                            RETURN
C
      END
C
C
C
      SUBROUTINE BCKSLV( A,NA,COL,N,RANK,X,Y )
C
      INTEGER NA,N,COL(N),RANK,I,J
      REAL*8 A(NA,N),X(N),Y(N),P
C
      DO   2000 I=RANK,1,-1
      P = Y(I)
      IF( I.EQ.N )                          GOTO  2000
      DO   1900 J = I+1,N
 1900 P = P - A(I,J)*Y(J)
 2000 Y(I) = P*A(I,I)
C
      DO   3000 I=1,N
 3000 X(COL(I)) = Y(I)
                                            RETURN
      END
```

6.13 Subroutine ORDER

ORDER is a very simple routine which rearranges the solution vector output from subroutine REDSOL into a form consistent with the rest of the formulation given by Equation (2.39). The resulting order vectors FIXBND and SOLV will contain the complete set of boundary quantities.

```
      SUBROUTINE ORDER(FIXBND,SOLV,KODE)
C
      IMPLICIT REAL*8(A-H,O-Z)
      COMMON/IO/INDEV,IOUTDV
      COMMON/MODEL/NG1,NTOT,NTOT2,NUMNP,NUMELE,NUMEQ,NUMINT,ITYPE,IPLOT
      DIMENSION FIXBND(1),SOLV(1),KODE(1)
C
C     ==========================================
C     =   REARRANGE SOLV AND FIXBND FOR INTERNAL   =
C     =   POINT CALCULATIONS                        =
C     ==========================================
C
      DO 10 J=1,NUMNP
C
         JJ=J+NUMNP
         IF(KODE(J).EQ.12)GOTO 20
         IF(KODE(J).EQ.13)GOTO 10
         IF(KODE(J).EQ.14)GOTO 30
         IF(KODE(J).EQ.23)GOTO 40
         IF(KODE(J).EQ.24)GOTO 50
C
      WRITE(IOUTDV,1000)
      RETURN
C
 20      SL=SOLV(J)
         SOLV(J)=FIXBND(JJ)
         FIXBND(JJ)=SL
         GOTO 10
C
 30      SL=SOLV(JJ)
         SOLV(JJ)=FIXBND(JJ)
         FIXBND(JJ)=SL
         GOTO 10
C
 40      SL=SOLV(J)
         SOLV(J)=FIXBND(J)
         FIXBND(J)=SL
         GOTO 10
C
 50      SL=SOLV(J)
         SL1=SOLV(JJ)
         SOLV(J)=FIXBND(J)
         SOLV(JJ)=FIXBND(JJ)
         FIXBND(J)=SL
         FIXBND(JJ)=SL1
C
 10   CONTINUE
C
      RETURN
C
 1000 FORMAT (//' *** CODE NUMBER ERROR *** '//)
C
      END
```

6.14 Subroutine TERNAL

Once a complete set of boundary information is determined by REDSOL and appropriately collocated by ORDER, this subroutine will compute the value of the field function, the Laplacian of the field function and their derivatives in both coordinate directions. The formulation used in calculating the two field variables is derived from Equations (2.11) and (2.12). The derivative evaluations are based on Equations (2.15), (2.16), (2.21), and (2.22). The reader is also referred to Section 4.4 of this text for more details of the formulation.

The structure of this routine is designed to distinguish between the four different element types and call the appropriate integration subroutine for the particular element under consideration. The influence of both the boundary quantities and the source term $f(x,y)$ on the calculation of internal point values are evaluated within this subroutine.

```
      SUBROUTINE TERNAL(FIXBND,SOLV,KODE,XP,YP,X,Y,S1,S2,
     $                  DS1DX,DS1DY,DS2DX,DS2DY,NELE)
C
C  ================================================================
C  =                                                              =
C  =   THIS SUBROUTINE COMPUTES THE POTENTIAL VALUES AT           =
C  =   THE INTERNAL POINTS SPECIFIED BY THE USER.                 =
C  =                                                              =
C  ================================================================
C
      IMPLICIT REAL*8(A-H,O-Z)
      REAL*8 KI,KJ,KK,KL,LI,LJ,LK,LL,KIX,KJX,KKX,KLX,LIX,LJX,LKX,LLX,
     &       KIY,KJY,KKY,KLY,LIY,LJY,LKY,LLY
      COMMON/IO/INDEV,IOUTDV
      COMMON/MODEL/NG1,NTOT,NTOT2,NUMNP,NUMELE,NUMEQ,NUMINT,ITYPE,IPLOT
      COMMON/SOURCE/NSOUR,SLX(10),SLY(10),SVALUE(10)
      DIMENSION FIXBND(1),SOLV(1),X(1),Y(1),KODE(1),NELE(NTOT,4)
C
      DATA OO2PI/0.159154943/
      DATA SMALL/1.0E-20/
C
C  =======================================
C  =  COMPUTATION OF INTERNAL POTENTIALS =
C  =======================================
C
            BT1=0.0
            BT2=0.0
            BT1X=0.0
            BT1Y=0.0
            BT2X=0.0
            BT2Y=0.0
            S1=0.0
```

```
            S2=0.0
            DS1DX=0.0
            DS1DY=0.0
            DS2DX=0.0
            DS2DY=0.0
C
        DO 200 JJ=1,NUMELE
C
            N1=NELE(JJ,1)
            N2=NELE(JJ,2)
            N3=NELE(JJ,3)
            N4=NELE(JJ,4)
            N11=N1+NUMNP
            N22=N2+NUMNP
            N33=N3+NUMNP
            N44=N4+NUMNP
            XI=X(N1)
            YI=Y(N1)
            XJ=X(N2)
            YJ=Y(N2)
C
        IF(N3.LE.0)THEN
C
            CALL INTEG(XP,YP,N1,N2,N3,GI,GJ,HI,HJ,KI,KJ,LI,LJ,
     $                              ITYPE,B1,B2,X,Y)
C
            S1=S1+FIXBND(N1)*HI+FIXBND(N2)*HJ
     $              -SOLV(N1)*GI-SOLV(N2)*GJ+FIXBND(N11)*LI
     $              +FIXBND(N22)*LJ-SOLV(N11)*KI-SOLV(N22)*KJ
            S2=S2+FIXBND(N11)*HI+FIXBND(N22)*HJ
     $              -SOLV(N11)*GI-SOLV(N22)*GJ
C
            BT1=BT1+B1
            BT2=BT2+B2
C
            CALL INTEGD(XP,YP,N1,N2,N3,GIX,GJX,HIX,HJX,KIX,KJX,
     $                              LIX,LJX,ITYPE,B1X,B2X,1,X,Y)
C
            DS1DX=DS1DX+FIXBND(N1)*HIX+FIXBND(N2)*HJX
     $              -SOLV(N1)*GIX-SOLV(N2)*GJX+FIXBND(N11)*LIX
     $              +FIXBND(N22)*LJX-SOLV(N11)*KIX-SOLV(N22)*KJX
            DS2DX=DS2DX+FIXBND(N11)*HIX+FIXBND(N22)*HJX
     $              -SOLV(N11)*GIX-SOLV(N22)*GJX
C
            BT1X=BT1X+B1X
            BT2X=BT2X+B2X
C
            CALL INTEGD(XP,YP,N1,N2,N3,GIY,GJY,HIY,HJY,KIY,KJY,
     $                              LIY,LJY,ITYPE,B1Y,B2Y,2,X,Y)
C
            DS1DY=DS1DY+FIXBND(N1)*HIY+FIXBND(N2)*HJY
     $              -SOLV(N1)*GIY-SOLV(N2)*GJY+FIXBND(N11)*LIY
     $              +FIXBND(N22)*LJY-SOLV(N11)*KIY-SOLV(N22)*KJY
            DS2DY=DS2DY+FIXBND(N11)*HIY+FIXBND(N22)*HJY
     $              -SOLV(N11)*GIY-SOLV(N22)*GJY
C
            BT1Y=BT1Y+B1Y
            BT2Y=BT2Y+B2Y
C
        ELSE
C
          IF(N4.EQ.0)THEN
C
            CALL INTEGQ(X,Y,XP,YP,N1,N2,N3,GI,GJ,GK,HI,HJ,HK,
     $                      KI,KJ,KK,LI,LJ,LK,ITYPE,B1,B2)
C
            S1=S1+FIXBND(N1)*HI+FIXBND(N2)*HJ+FIXBND(N3)*HK
     $              -SOLV(N1)*GI-SOLV(N2)*GJ-SOLV(N3)*GK
     $              +FIXBND(N11)*LI+FIXBND(N22)*LJ+FIXBND(N33)*LK
```

```
     $               -SOLV(N11)*KI-SOLV(N22)*KJ-SOLV(N33)*KK
               S2=S2+FIXBND(N11)*HI+FIXBND(N22)*HJ+FIXBND(N33)*HK
     $               -SOLV(N11)*GI-SOLV(N22)*GJ-SOLV(N33)*GK
C
               BT1=BT1+B1
               BT2=BT2+B2
C
               CALL INTGQD(X,Y,XP,YP,N1,N2,N3,GIX,GJX,GKX,HIX,HJX,HKX,
     $                    KIX,KJX,KKX,LIX,LJX,LKX,ITYPE,B1X,B2X,1)
C
               DS1DX=DS1DX+FIXBND(N1)*HIX+FIXBND(N2)*HJX+FIXBND(N3)*HKX
     $               -SOLV(N1)*GIX-SOLV(N2)*GJX-SOLV(N3)*GKX
     $               +FIXBND(N11)*LIX+FIXBND(N22)*LJX+FIXBND(N33)*LKX
     $               -SOLV(N11)*KIX-SOLV(N22)*KJX-SOLV(N33)*KKX
               DS2DX=DS2DX+FIXBND(N11)*HIX+FIXBND(N22)*HJX+FIXBND(N33)*HKX
     $               -SOLV(N11)*GIX-SOLV(N22)*GJX-SOLV(N33)*GKX
C
               BT1X=BT1X+B1X
               BT2X=BT2X+B2X
C
               CALL INTGQD(X,Y,XP,YP,N1,N2,N3,GIY,GJY,GKY,HIY,HJY,HKY,
     $                    KIY,KJY,KKY,LIY,LJY,LKY,ITYPE,B1Y,B2Y,2)
C
               DS1DY=DS1DY+FIXBND(N1)*HIY+FIXBND(N2)*HJY+FIXBND(N3)*HKY
     $               -SOLV(N1)*GIY-SOLV(N2)*GJY-SOLV(N3)*GKY
     $               +FIXBND(N11)*LIY+FIXBND(N22)*LJY+FIXBND(N33)*LKY
     $               -SOLV(N11)*KIY-SOLV(N22)*KJY-SOLV(N33)*KKY
               DS2DY=DS2DY+FIXBND(N11)*HIY+FIXBND(N22)*HJY+FIXBND(N33)*HKY
     $               -SOLV(N11)*GIY-SOLV(N22)*GJY-SOLV(N33)*GKY
C
               BT1Y=BT1Y+B1Y
               BT2Y=BT2Y+B2Y
C
          ELSE
C
               CALL INTEGO(X,Y,XP,YP,N1,N2,N3,N4,GI,GJ,GK,GL,HI,HJ,HK,HL,
     $                    KI,KJ,KK,KL,LI,LJ,LK,LL,ITYPE,B1,B2)
C
       S1=S1+FIXBND(N1)*HI+FIXBND(N2)*HJ+FIXBND(N3)*HK+FIXBND(N4)*HL
     $    -SOLV(N1)*GI-SOLV(N2)*GJ-SOLV(N3)*GK-SOLV(N4)*GL
     $    +FIXBND(N11)*LI+FIXBND(N22)*LJ+FIXBND(N33)*LK+FIXBND(N44)*LL
     $    -SOLV(N11)*KI-SOLV(N22)*KJ-SOLV(N33)*KK-SOLV(N44)*KL
       S2=S2+FIXBND(N11)*HI+FIXBND(N22)*HJ+FIXBND(N33)*HK+FIXBND(N44)*HL
     $    -SOLV(N11)*GI-SOLV(N22)*GJ-SOLV(N33)*GK-SOLV(N44)*GL
C
               BT1=BT1+B1
               BT2=BT2+B2
C
               CALL INTGOD(X,Y,XP,YP,N1,N2,N3,N4,GIX,GJX,GKX,GLX,HIX,HJX,
     $          HKX,HLX,KIX,KJX,KKX,KLX,LIX,LJX,LKX,LLX,ITYPE,B1X,B2X,1)
C
       DS1DX=DS1DX+FIXBND(N1)*HIX+FIXBND(N2)*HJX+FIXBND(N3)*HKX
     $    +FIXBND(N4)*HLX-SOLV(N1)*GIX-SOLV(N2)*GJX-SOLV(N3)*GKX
     $    -SOLV(N4)*GLX+FIXBND(N11)*LIX+FIXBND(N22)*LJX+FIXBND(N33)*LKX
     $    +FIXBND(N44)*LLX-SOLV(N11)*KIX-SOLV(N22)*KJX-SOLV(N33)*KKX
     $    -SOLV(N44)*KLX
       DS2DX=DS2DX+FIXBND(N11)*HIX+FIXBND(N22)*HJX+FIXBND(N33)*HKX
     $       +FIXBND(N44)*HLX-SOLV(N11)*GIX-SOLV(N22)*GJX-SOLV(N33)*GKX
     $       -SOLV(N44)*GLX
C
               BT1X=BT1X+B1X
               BT2X=BT2X+B2X
C
C
               CALL INTGOD(X,Y,XP,YP,N1,N2,N3,N4,GIX,GJX,GKX,GLX,HIX,HJX,
     $          HKX,HLX,KIX,KJX,KKX,KLX,LIX,LJX,LKX,LLX,ITYPE,B1Y,B2Y,2)
C
       DS1DY=DS1DY+FIXBND(N1)*HIX+FIXBND(N2)*HJX+FIXBND(N3)*HKX
     $    +FIXBND(N4)*HLX-SOLV(N1)*GIX-SOLV(N2)*GJX-SOLV(N3)*GKX
```

```
      $     -SOLV(N4)*GLX+FIXBND(N11)*LIX+FIXBND(N22)*LJX+FIXBND(N33)*LKX
      $     +FIXBND(N44)*LLX-SOLV(N11)*KIX-SOLV(N22)*KJX-SOLV(N33)*KKX
      $     -SOLV(N44)*KLX
       DS2DY=DS2DY+FIXBND(N11)*HIX+FIXBND(N22)*HJX+FIXBND(N33)*HKX
      $        +FIXBND(N44)*HLX-SOLV(N11)*GIX-SOLV(N22)*GJX-SOLV(N33)*GKX
      $        -SOLV(N44)*GLX
C
             BT1Y=BT1Y+B1Y
             BT2Y=BT2Y+B2Y
C
           END IF
C
          END IF
C
 200      CONTINUE
C
          IF(ITYPE.EQ.3)THEN
             CALL FAN(XP,YP,X,Y,NELE,BT1,BT2,BT1X,BT1Y,BT2X,BT2Y)
          END IF
C
C ===================================
C =  POINT SOURCE TERM CALCULATION  =
C ===================================
C
      IF(NSOUR.GE.1)THEN
C
         DO 210 IS=1,NSOUR
C
         DXS=SLX(IS)-XP
         DYS=SLY(IS)-YP
         ARG1=DXS*DXS+DYS*DYS
         IF(ARG1.EQ.0.0)ARG1=SMALL
         ALNX=DLOG(ARG1)
C
         BT1=BT1-SVALUE(IS)*0.125*ARG1*(ALNX-2.0)
         BT2=BT2-SVALUE(IS)*0.5*ALNX
         BT1X=BT1X+SVALUE(IS)*0.25*DXS*(ALNX-1.0)
         BT1Y=BT1X+SVALUE(IS)*0.25*DYS*(ALNX-1.0)
         BT2X=BT2X+SVALUE(IS)*DXS/ARG1
         BT2Y=BT2Y+SVALUE(IS)*DYS/ARG1
C
 210     CONTINUE
C
      END IF
C
         S1=(S1+BT1)*(-002PI)
         S2=(S2+BT2)*(-002PI)
         DS1DX=(DS1DX+BT1X)*(-002PI)
         DS1DY=(DS1DY+BT1Y)*(-002PI)
         DS2DX=(DS2DX+BT2X)*(-002PI)
         DS2DY=(DS2DY+BT2Y)*(-002PI)
C
      RETURN
      END
```

6.15 Subroutine OUTPUT

OUTPUT is the final routine the program executed. This subroutine serves
two functions. First, it coordinates the calculation of internal point values.
Secondly, as its name implies, it outputs the results of the boundary element an-

alysis in terms of both the four boundary quantities and any specified internal point values.

```
      SUBROUTINE OUTPUT(X,Y,SOLV,FIXBND,XCENT,YCENT,NELE,KODE)
C
C     ============================================================================
C     =                                                                          =
C     =   THIS SUBROUTINE WRITES OUT THE RESULTS OF THE COMPUTATION               =
C     =                                                                          =
C     =      CALLING ARGUMENTS -                                                  =
C     =                                                                          =
C     =      X,Y --------- VECTORS OF BOUNDARY COORDINATES                        =
C     =      FIXBND ------ VECTOR CONTAINING POTENTIALS AT NODES                  =
C     =      SOLV -------- VECTOR CONTAINING FLUXES AT NODES                      =
C     =      XCENT,YCENT - VECTORS CONTAINING INTERNAL COORDINATES                =
C     =                                                                          =
C     ============================================================================
C
      IMPLICIT REAL*8(A-H,O-Z)
      DIMENSION X(1),Y(1),FIXBND(1),SOLV(1),XCENT(1),YCENT(1),
     &          SOLUT1(100),SOLUT2(100),DF1DX(100),DF1DY(100),
     &          DF2DX(100),DF2DY(100),NELE(NTOT,4),KODE(1)
      COMMON/IO/INDEV,IOUTDV
      COMMON/MODEL/NG1,NTOT,NTOT2,NUMNP,NUMELE,NUMEQ,NUMINT,ITYPE,IPLOT
C
      CALL ORDER(FIXBND,SOLV,KODE)
C
      WRITE(IOUTDV,100)
C
      DO 10 I=1,NUMNP
      II=I+NUMNP
         WRITE(IOUTDV,200)I,FIXBND(I),SOLV(I),FIXBND(II),SOLV(II)
  10  CONTINUE
C
      IF(NUMINT.EQ.O)RETURN
C
C     ================================================
C     =       COMPUTATION OF INTERNAL POTENTIALS      =
C     ================================================
C
      WRITE(7,150)
C
      DO 15 NP=1,NUMINT
C
      XP=XCENT(NP)
      YP=YCENT(NP)
C
      CALL TERNAL(FIXBND,SOLV,KODE,XP,YP,X,Y,S1,S2,
     $                 S1X,S1Y,S2X,S2Y,NELE)
C
       SOLUT1(NP)=S1
       SOLUT2(NP)=S2
       DF1DX(NP)=S1X
       DF2DX(NP)=S2X
       DF1DY(NP)=S1Y
       DF2DY(NP)=S2Y
C
  15  CONTINUE
C
      WRITE(IOUTDV,300)
      DO 20 I=1,NUMINT
         WRITE(IOUTDV,400)I,SOLUT1(I),DF1DX(I),DF1DY(I)
  20  CONTINUE
```

```
c
      WRITE(IOUTDV,310)
      DO 30 I=1,NUMINT
         WRITE(IOUTDV,400)I,SOLUT2(I),DF2DX(I),DF2DY(I)
 30   CONTINUE
c
      RETURN
c
c   ====================
c   = FORMAT STATEMENTS =
c   ====================
c
 100  FORMAT (//80('-')//35X,'SOLUTION'//80('-')//1X,'NODE',10X,
      $         'F',15X,'F¢',15X,'L(F)',15X,'L(F)¢'//)
 150  FORMAT(1X,' CALCULATING INTERNAL POINTS')
 200  FORMAT (I4,4E18.8)
 300  FORMAT (//80('-')//30X,'INTERNAL POINTS'//80('-')//1X,
      $         'INTERNAL POINT',5X,'F',17X,'dF/dx',16X,'dF/dy'//)
 310  FORMAT (//80('-')//1X,'INTERNAL POINT',3X,'L(F)',14X,'dL(F)/dx',
      $         13X,'dL(F)/dy'//)
 400  FORMAT(5X,I3,3E20.8)
c
      END
```

6.16 Functions F, FN, FL, and FLN

These user-dependent functions serve to define the form of the nonhomogeneous source term f(x,y). If the source term f(x,y) is a biharmonic function, then
the user simply writes, using standard FORTRAN conventions, the definition of
f(x,y) in FUNCTION F, the normal derivative of f(x,y) in FUNCTION FN, the Laplacian of f(x,y) in FUNCTION FL, and finally the normal derivative of the Laplacian
of f(x,y) in FUNCTION FLN. If any of the derivatives of f(x,y) are zero then the
corresponding function statement must also be defined as zero.

However, if the source term f(x,y) is of an arbitrary form such that subroutine FAN is used to evaluate the domain integrations, only FUNCTION F need be
defined. The derivative forms of the f(x,y) are not called by the domain "fanning" routine.

```
c
c---------------------------------------------------- SOURCE TERM FUNCTION
c
      FUNCTION F(X,Y)
c
c   ======================================================================
c   = THIS FUNCTION COMPUTES THE NON-HOMOGENEOUS PORTION OF THE          =
c   = BIHARMONIC EQUATION AT A POINT AND IS CHANGEABLE BY THE USER.      =
c   =                                                                    =
c   =     CALLING ARGUMENTS -                                            =
c   =         X,Y - COORDINATES OF POINT IN QUESTION                     =
c   ======================================================================
```

```
C
      IMPLICIT REAL*8(A-H,O-Z)
C
        F=X**3
C
      RETURN
      END
C
C-------------------------------- NORMAL DERIVATIVE OF HARMONIC
C                                 SOURCE TERM
C
      FUNCTION FN(X,Y,DCX,DCY)
C
C =====================================================================
C = THIS FUNCTION COMPUTES THE NORMAL DERIVATIVE OF NON-HOMOGENEOUS  =
C = PORTION OF THE BIHARMONIC EQUATION AT A POINT AND IS             =
C = CHANGEABLE BY THE USER. (NOT ALWAYS NECESSARY)                   =
C =                                                                  =
C =     CALLING ARGUMENTS -                                          =
C =        X,Y - COORDINATES OF POINT IN QUESTION                    =
C =====================================================================
C
      IMPLICIT REAL*8(A-H,O-Z)
C
        FN=3.O*X*X*DCX
C
      RETURN
      END
C
C-------------------------------- LAPLACIAN OF THE SOURCE TERM FUNCTION
C
      FUNCTION FL(X,Y)
C
C =====================================================================
C = THIS FUNCTION COMPUTES THE LAPLACIAN OF THE NON-HOMOGENEOUS      =
C = PORTION OF THE BIHARMONIC EQUATION AT A POINT AND IS CHANGEABLE  =
C = BY THE USER. (NOT ALWAYS NECESSARY)                             =
C =                                                                  =
C =     CALLING ARGUMENTS -                                          =
C =        X,Y - COORDINATES OF POINT IN QUESTION                    =
C =====================================================================
C
      IMPLICIT REAL*8(A-H,O-Z)
C
        FL=6.O*X
C
      RETURN
      END
C
C-------------------------------- NORMAL DERIVATIVE OF THE LAPLACIAN
C                                 OF THE SOURCE TERM
C
      FUNCTION FLN(X,Y,DCX,DCY)
C
C =====================================================================
C = THIS FUNCTION COMPUTES THE NORMAL DERIVATIVE OF THE LAPLACIAN OF =
C = NON-HOMOGENEOUS PORTION OF THE BIHARMONIC EQUATION AT A POINT    =
C = AND IS CHANGEABLE BY THE USER. (NOT ALWAYS NECESSARY)            =
C =                                                                  =
C =     CALLING ARGUMENTS -                                          =
C =        X,Y - COORDINATES OF POINT IN QUESTION                    =
C =====================================================================
C
      IMPLICIT REAL*8(A-H,O-Z)
C
        FLN=6.O*DCX
C
      RETURN
      END
```

6.17   Input Data Structure

In this section a complete description of how to define an appropriate input data set to model a biharmonic problem is presented. The user should find all the necessary information to construct a boundary element approximation to any appropriate biharmonic problem. An example input data set will be presented after the input structure is defined.   The terms in parenthesis indicate the format of each input line.

A.   Problem Identification - (20A4)

Columns 1 to 80 of the first input line contain information the user specifies to be printed with the output of the results.

B.   Control Parameters - (5I5)

Columns 1 - 5   Problem type

= 0 for homogeneous analysis

= 1 for exact analysis for harmonic source term

for rectilinear geometries only

= 2 for biharmonic source term analysis

= 3 for arbitrary source analysis

6 - 10   Number of nodal points (200 maximum)

11 - 15   Number of elements

16 - 20   Number of internal points where the solution is desired

21 - 25   Number of internal point sources

C.   Point Source Information - (3F10.3)

If the number of internal source points is non-zero then this line is read as input.   However, if the number of internal source points is zero this line is omitted.

```
Columns   1 - 10   x-coordinate of source point
         11 - 20   y-coordinate of source point
         21 - 30   Source point strength
```

D. Special Harmonic Source Term Definition - (4F10.2)

If the problem type is specified as one, this input line must be defined,
otherwise it is omitted.   The special harmonic form of f(x,y) is defined by:
f(x,y) = C1xy + C2x + C3y + C4.

```
Columns  1 - 10   Coefficient C1
        11 - 20   Coefficient C2
        21 - 30   Coefficient C3
        31 - 40   Coefficient C4
```

E. Nodal Point Definition - (2I5,4F15.0)

```
Columns   1 -  5   Nodal point number
          6 - 10   Boundary condition code
                   = 1 if the function is specified at node
                   = 2 if the normal derivative of the function is
                         specified at node
                   = 3 if the Laplacian of function is specified at node
                   = 4 if the normal derivative of the Laplacian of function
                         is specified at node
*** NOTE ***       A two digit code is necessary since for a well-posed
                   boundary value problem two boundary conditions must be
                   specified at each nodal point.  Possible code specifica-
                   tions: 12, 13, 14, 23, 24
         11 - 25   x-coordinate of node
         26 - 40   y-coordinate of node
```

41 - 55   Boundary values at node associated with first digit
of boundary condition code

56 - 70   Boundary values at node associated with second digit
of boundary condition code

Nodal point input lines should be in number in a sequential order starting with the one. If any node numbers are omitted in the sequence, they will be automatically generated at equal intervals along a straight line between the specified nodal points. The specified boundary values will also be distributed linearly over the generated section. However, the value of the boundary code is set equal to the first line in the generated sequence.

Two nodes may occupy the same coordinate space if the "double noding" technique described in Chapter 4 is used. However, they must have different boundary conditions codes for the solution algorithm is function successfully.

F. Internal Point Definition - (I10,2F15.0)

If the number of internal points where a solution is desired is zero, this set of input lines should be omitted. However, if internal point values are specified by the user, they should be numbered sequentially starting from one. Any omitted points will be generated in a straight line between the two specified points.

Columns 1 - 10   Internal point number

11 - 25   x-coordinate of internal point

26 - 40   y-coordinate of internal point

G. Element Definition - (5I5)

Elements are prescribed by denoting the boundary node numbers defining that element. The elements are numbered sequentially in a counterclockwise order around the periphery of the domain (or clockwise for an external problem). The

node numbers N1, N2, N3, and N4 allow the user to specify any one of four element
types.

    Column 1 - 5   Element number
            6 - 10   Node N1
           11 - 15   Node N2
           16 - 20   Node N3
           21 - 25   Node N4

    Element types: Constant - N1 and N2 are the end nodes, N3 must be
                               negative, and N4 = 0. The actual discrete
                               boundary node is located at the center of
                               the element
                   Linear   - N1 and N2 are the end nodes, N3=0, N4=0
                   Quadratic - N1, N2, and N3 are the three nodes defining
                               the element, N4 = 0
                   Overhauser - N1, N2, N3, and N4 are the four nodes
                               defining the element

Omitted element input lines are generated consistent with the first and last
line of the generation sequence. If the "double noding" technique described in
Chapter 4 is used, each node constituting the double node set must exist in one
element only. In other words, no element can be defined with both nodes of the
double node set. (The result is an element of zero length)

6.18 Sample Input Data

In this section an example input data set based on the problem defined in
Figure 6.1 will be presented to demonstrate the basic input structure and illustrate some of the node and element generating capabilities. The number next to
each node is the node number and the circled numbers are the element numbers.
It is instructive to note, for users unfamiliar with the technique, how each
double noded corner is handled in the input data set.

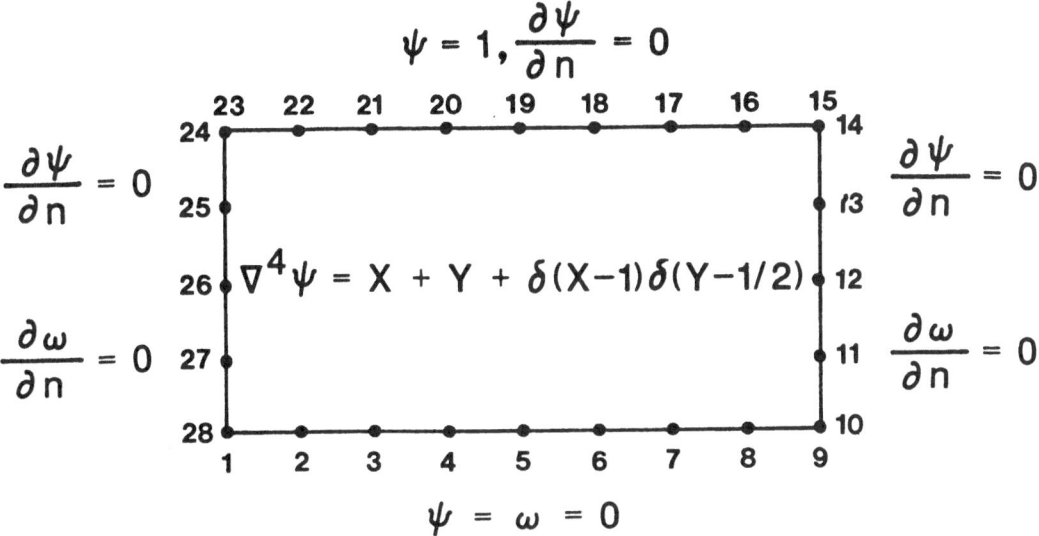

$$\psi = 1, \frac{\partial \psi}{\partial n} = 0$$

$$\frac{\partial \psi}{\partial n} = 0$$

$$\frac{\partial \psi}{\partial n} = 0$$

$$\nabla^4 \psi = X + Y + \delta(X-1)\delta(Y-1/2)$$

$$\frac{\partial \omega}{\partial n} = 0$$

$$\frac{\partial \omega}{\partial n} = 0$$

$$\psi = \omega = 0$$

```
Example Problem - Harmonic Source Term f(x,y)=x+y With One Point Source
  1    28    20     4     1
1.0          0.5          1.0
0.0          1.0          1.0          0.0
  1   13      0.0         0.0          0.0          0.0
  9   13      2.0         0.0          0.0          0.0
 10   24      2.0         0.0          0.0          0.0
 14   24      2.0         2.0          0.0          0.0
 15   12      2.0         2.0          1.0          0.0
 23   12      0.0         2.0          1.0          0.0
 24   13      0.0         2.0          0.0          0.0
 28   13      0.0         0.0          0.0          0.0
       1      0.2         0.5
       4      0.8         0.5
  1    1     2
  2    2     3
  3    2     3     4     5
  6    5     6     7     8
  7    7     8
  8    8     9
  9   10    11
 10   11    12    13
 11   13    14
 12   15    16    17
 13   16    17    18    19
 16   19    20    21    22
 17   21    22    23
 18   24    25
 19   25    26    27
 20   27    28
```

Figure 6.1   Example Problem and Corresponding Input Data Set

CHAPTER 7

GENERAL SUMMARY AND CONCLUDING REMARKS

7.1 Summary

The objective of this text was to present a more-or-less self-contained treatment of boundary integral analysis of the nonhomogeneous biharmonic equation, and to present certain refinements that make the method more accurate as well as computationally more efficient. The necessary boundary integral relationships for very general cases have been derived. These include the incorporation of nonhomogeneous terms as "body force" type domain integrals and nonlinear terms as "pseudo-body force" domain integrals.

Detailed analyses and implementation strategies for various element types have been described. The philosophy of taking the analytical formulation to the extreme before going to the computer has been invoked throughout the text. In addition to the traditional element types that have been used in boundary element surface discretization, an improved element, the "Overhauser" spline, has been introduced as a superior performer for use in biharmonic analysis. This is attributed to the Overhauser's improved ability to represent accurately the domain boundary and the corresponding variation of the field variable in configuration space.

In the way of progress in handling domain integrations, a "fanning" integration technique has been formalized into an algorithm that is simply implemented. The integration scheme is characterized by an implicit domain discretization (one which does not require analyst pre-preparation or intervention) that accurately integrates both the singular and nonsingular integrands occurring in the biharmonic formulation.

With the Overhauser element and the domain fanning system, a wide variety of problems have been solved with a precision and ease that is unprecedented in previous work. These problems include both nonhomogeneous and nonlinear variations of the basic biharmonic problem. The nonlinear problems were handled by incorporating an iterative solution technique which takes advantage of an interpolating domain map storage scheme. Precision solutions have been obtained for a diverse group of problems in which the nonlinearities involved both the field variable and its various derivatives. The example problems, presented in Chapter 5, not only corroborate this claim of accuracy and versatility but also imply the formulation's ability to solve similar problems of this nature.

Lastly, a detailed computer program suitable for solving a general two-dimensional nonhomogeneous biharmonic problem has been presented. The program includes constant, linear, quadratic, and Overhauser elements, as well as the fanning domain integrator for handling general domain terms. This program may be used to solve any of the single iteration problems described in Chapter 5.

7.2 Conclusions

Several conclusions are evident from the research portion of this work and it is easy to extrapolate these into general statements that carry over into other areas of BEM. Primary among these issues is the piecewise representation of the bounding surface that characterizes "boundary element" methods generally. This is quite naturally a more fundamental problem in BEM than in finite elements or finite differences, which are both domain-based methods. Since ideally, the boundary integral representation will use discrete boundary information exclusively in order to effect a solution, it would seem reasonable to assume that BEM would be more sensitive to boundary representation error than either of the aforementioned domain-based techniques.

Although this conjecture has not been proven rigorously, the results from the example problems indicate progressively better solution accuracy results as the element representation of the boundary becomes better. Continuing to speak empirically, the reasons for this are not difficult to rationalize. Even when, say a rectangular plate, is modelled geometrically precisely with linear elements, the displacement field is not likely to be represented accurately with such a formulation since the displaced configuration is not likely to be linear. The problem is compounded when domain integrations are involved, and is especially evident when the nonhomogeneous functions are themselves 2n-harmonic. In these few important cases where the domain integrals may be transformed into equivalent surface integrals, the domain effect is transformed into what may be an inaccurate representation of the geometrical boundary. Thus, an avoidable inaccuracy is introduced when using the simpler elements.

Because of its simple formulation, its localized C_1 continuity, and its generally superior results, the Overhauser spline element is concluded to be an all-around superior element. The only disadvantage of using the Overhauser is in modelling surfaces which should not be splined. Such surfaces are characterized by the presence of true derivative discontinuous geometries, i.e., a corner of a plate. For this reason, no general analysis program should contain Overhauser elements alone, but should include simpler elements in order to handle C_1 discontinuous surfaces. Thus, a mesh with a predominant presence of Overhauser elements, but with lower order elements at corners is concluded to give the best results.

In regard to handling discontinuous or ambiguous boundary conditions at certain locations (typically at corners), three strategies may be employed. The first is to use discontinuous elements at that location; that is, use, say, a constant element on either side of the troublesome point. Since there is no node there, the problem is empirically circumvented. The second technique is to double-node the point in the manner advocated by Brebbia (1978). This method involves placing two nodes "very close" to the offensive point and linking them with an extremely short element several orders of magnitude shorter than any others appearing in the mesh. The third method involves double-noding again, but without physically linking the elements containing the two nodes. The nodes actually reside at the same point (the trouble spot) but on different elements with different boundary conditions. An actual physical break in the continuity of the mesh occurs at this point. Each node assumes its own position in displacement space. This technique, which is actually a compromise between the first two methods, has consistently given superior results in our tests, and is, therefore, the recommended method to use.

Regardless of the element type used, error may be introduced in the basic assembly of the matrices due to inaccurate numerical integrations. In this work, we have advocated performing as much as possible of the integration analytically before resorting to the numerical quadrature. (This is in stark contrast to finite elements where the opposite view is routinely taken for efficiency reasons.) A series of exact analytical expressions for the required surface integrations were developed for general isoparametric linear elements and the subparametric forms of the quadratic and Overhauser elements. A hybrid numerical/analytical scheme is adopted for the isoparametric forms of the quadratic and Overhauser elements.

The subparametric versions of both the quadratic and the Overhauser elements have been found to perform excellently for rectilinear geometries. These elements apparently effect a compromise between linear elements for efficiency of modelling linear geometries, and the higher order elements for representing displacements in configuration space. A significant reduction in computer solution time results when these elements are used instead of their isoparametric counterparts. The superiority of the Overhauser element over the quadratic element continues to exist even in this subparameterized format, due to this latter element's inability to maintain C_1 interelement continuity.

As to evaluating domain integrals in the nonhomogeneous form of the biharmonic equation, conclusions equally as significant in all physical applications of BEM are evident from this research work. The most common approach by far in past work for evaluating domain integrals has been to subdivide the domain into a series of cells over which some type of numerical quadrature is performed. However, this approach naturally involves some type of meshing system similar to that used in finite elements, and, hence, removes one of BEM's most practical advantages. As a matter of principle, the research group, headed by the second author, attempts to avoid this expedient at every opportunity since it is not in the "spirit" of BEM. Consequently, several other techniques are advocated as superior tactical substitutes. The Monte Carlo quadrature technique (Gipson, 1985) is quite general and a very powerful method for handling the most general domain integrals, without sacrificing utility. However, the method does suffer from needing a relatively large number of computations to obtain sufficient accuracy, and from a lack of uniform convergence as the number of quadrature points is increased. Several methods involving multi-dimensional Gaussian quadrature without explicit gridding are possible alternatives if the behavior of the integrated function is known a priori. If the function varies rapidly at any particular point, the integral may be subdivided into a set of smaller subregions over which the function is less radically behaved. The "fanning" domain integration method formalized in this work draws on the advantages of implicit domain integration while automatically concentrating quadrature points in a way that is sensitive to the possible singular nature of the integrand. In the particular method developed in this work, the number of quadrature points is directly related to the number of surface elements used to define the boundary of the problem.

The most effective and accurate method for effecting domain integrations is through the use of integral transformations when the function types are amenable to this technique. In this method, Green's identity is used to transform the domain integral into an equivalent surface integral whenever the functions are "2n-harmonic" (Gipson and Reible, 1986; Gipson et al., 1987). The power and accuracy of the higher order elements, especially the Overhauser, make this technique extremely attractive. The drawback of the method, the fact that the nonhomogeneous function must satisfy $\nabla^{2n}f = 0$, is not as severe as one might initially think; a large number of functions fall into this category. A physical example of such a contingency includes a uniform or linearly varying loading on a plate. Also, general transcendental functions may be represented by a finite series approximation and transformed by the appropriate form of Green's identity. This method was used to obtain very accurate results with relatively coarse boundary discretizations using Overhauser elements.

The other major domain term contingency, that of concentrated point loads, may be handled precisely with BEM. This technique, involving the representation of concentrated source in terms of the Dirac delta, eliminates any explicit integration whatsoever (Brebbia et al., 1984).

Significant progress was made in analyzing nonlinear forms of equations involving biharmonic operators. The iterative solution technique employing the interpolation map storage scheme proved to be quite efficient in solving the form of the equation in which the nonhomogeneous term was a function of the field variables and their various partial derivatives. A variety of examples presented in Chapter 5 demonstrated the ability of this method for solving extremely complex plate bending and creeping flow problems for which solutions by analytical means are virtually intractable. The interpolating nature of the map storage formulation reduces the amount of time required to update each map while providing an accurate representation of the solution over the domain. Each map is automatically generated and updated for regions of arbitrary shape without requiring information additional to that necessary to define the discrete boundary of the problem. As with the recommended domain discretization techniques, such a procedure shifts the burden of "domain bureaucracy" from the analyst to the computer, which is rightfully where the bulk of the work should be relegated.

The computer program presented is probably the only published one in existence as of this writing for the analysis of the general biharmonic equation.

The variety of element types available, as well as the sophisticated domain integration strategies available, make the program the most powerful reported to date for performing a general biharmonic analysis in two-dimensions.

7.3 Concluding Remarks

We should like to conclude this work with a few remarks in the way of editorial comments. The art and science of boundary elements is no longer in its infancy, but it is still in its childhood. Brebbia mentioned in the preface to Boundary Elements VII (1985) that the boundary element research community should learn from the mistakes of the early finite element researchers, and attempt to not go too far afield without quantifying the problems and errors associated with the present state of the art.

This is a wise philosophy. In order for the method to grow properly, it is necessary that any and all BEM techniques presented here or elsewhere be subjected to the utmost scrutiny. That is why we look forward to the present work being made obsolete, and replaced in the foreseeable future, if not by ourselves, by someone who has been sufficiently inspired and interested to take the work to further heights. Further, we hope that the interested reader has seen merit in this work besides in exclusive applications to the biharmonic equation, and will find the element types and domain integration techniques useful in other BEM analyses.

REFERENCES

Abramowitz, M., and Stegun, I. A., eds. Handbook of Mathematical Functions.
    Dover Publications, Inc., New York, 1972.

Altiero, N. J., and Sikarskie, D. L. "A Boundary Integral Method Applied to
    Plates of Arbitrary Plan Form." Computers and Structures, Vol. 9, 1978, pp.
    163-168.

Banerjee, P. K., and Butterfield, R. Boundary Element Methods in Engineering
    Science. McGraw-Hill, New York, 1981.

Bezine, G., and Gamby, D. "A New Integral Equation Formulation for Plate Bending
    Problems." Recent Advances in Boundary Element Methods. C.A. Brebbia,
    ed. Pentech Press, London, 1978.

Beyer, W. H., ed. CRC Standard Mathematical Tables. CRC Press, Inc., Boca
    Raton, Florida, 1981.

Bois, G. P. Tables of Indefinite Integrals. Dover Publications, Inc., New York,
    1961.

Brebbia, C. A., ed. The Boundary Element Method for Engineers. Pentech Press,
    London, 1978.

Brebbia, C. A., and Walker S., Boundary Element Techniques in Engineering.
    Newnes-Butterworths, London, 1980.

Brebbia, C. A., ed. Recent Advances in Boundary Element Methods, Proc. of the
    First Intl. Conference on Boundary Element Methods. Southampton University,
    CML Publications, London, 1980.

Brebbia, C. A., ed. New Developments in Boundary Element Methods, Proc. of the
    Second Intl. Conference on Boundary Element Methods. Southampton University,
    CML Publications, London, 1980.

Brebbia, C. A., ed. Boundary Element Methods, Proc. of the Third Intl. Confer-
    ence on Boundary Element Methods. Irvine, California, Springer-Verlag,
    Berlin, 1981.

Brebbia, C. A., ed. Boundary Element Methods in Engineering, Proc. of the Fourth
    Int. Conference on Boundary Element Methods. Southampton University,
    Springer-Verlag, Berlin, 1982.

Brebbia, C. A., Futagami, T., and Tanaka M., eds. Boundary Elements, Proc. of
    the Fifth Intl. Conference on Boundary Element Methods. Hiroshima, Japan,
    Springer-Verlag, Berlin, 1983.

Brebbia, C. A., ed. Progress in Boundary Element Methods. Vol. 2. Pentech
    Press, London, 1983.

Brebbia, C. A., Telles, J. C. F., and Wrobel, L. C. Boundary Element Techniques. Theory and Applications in Engineering. Springer-Verlag, Berlin, 1984.

Brebbia. C. A., ed. Boundary Element Techniques in Computer-Aided Engineering. Martinus Nijhoff Publishers, Dordrecht, 1984.

Brebbia, C. A., ed. Topics in Boundary Element Research. Springer-Verlag, Berlin, 1984.

Brebbia, C. A., and Maier, G., ed. Boundary Elements VII, Proc. of the Seventh Intl. Conference on Boundary Element Methods. Lake Como, Italy, 1985; Springer-Verlag, Berlin, 1985.

Brewer, J. A. "Three-Dimensional Design by Graphical Man-Computer Communication." Ph.D. dissertation, Purdue University, 1977.

Brewer, J. A., and Anderson, D. C.. "Visual Interaction with Over-hauser Curves and Surfaces." Computer Graphics, Vol. 11, 1977, pp. 132-137.

Butkovskiy, A. G. Green's Functions and Transfer Functions Hand-book. L. W. London, trans. Elliss Horwood Ltd., New York, 1982.

Camp, C. V., and Gipson, G. S. "A Boundary Element Method for Viscous Flows at Low Reynolds Number." Boundary Elements IX, Vol. 3,1987, pp. 419-432.

Camp, C. V. "A Solution to the Nonhomogeneous Biharmonic Equation by the Boundary Element Method." Ph.D. disseration, Oklahoma State University, 1987.

Connor, J. J., and Brebbia, C. A., eds. Betech 86: Proc. of the 2nd Boundary Element Technology Conference. Massachusetts Institute of Technology, June, 1986; Computational Mechanics Publications, Southampton, 1986.

Costa, J. A., and Brebbia C. A. "The Boundary Element Method Applied to Plates on Elastic Foundations." Engineering Analysis, Vol. 2, No. 4, 1985, pp. 174-183.

Cowper, G. R. "Gaussian Quadrature Formulas for Triangles." International Journal of Numerical Methods in Engineering, Vol. 7, 1973, pp. 405-408.

Currie, I. G. Fundamental Mechanics of Fluids. McGraw-Hill, New York, 1974.

Cushing, J. T. Applied Analytical Mathematics for Physical Scientists. John Wiley and Sons, New York, 1975.

Fairweather, G., Rizzo, F. J., Shippy, D. J., and Wu, Y. S. "On the Numerical Solution of Two-Dimensional Potential Problems by an Improved Boundary Integral Equation Method " Journal of Computational Physics, Vol. 31, 1979, pp. 96-112.

Gipson, G. S. "The Coupling of Monte Carlo Integration with the Boundary Integral Equation Technique to Solve Poisson Type Equations." Ph.D. dissertation, Louisiana State University, 1982.

Gipson, G. S. "Coupling Monte Carlo Quadrature with Boundary Elements to Handle Domain Integrals in Poisson Type Problems." Engineering Analysis, Vol. 2, No. 3, 1985, pp. 138-145.

Gipson, G. S., and Camp, C. V. "Effective Use of Monte Carlo Quadrature for Body Force Integrals Occurring in the Integral Form of Elastostatics." Boundary Elements VII: Proc. of the Seventh Intl. Conference on Boundary Element Methods. September, 1985, Lake Como, Italy, Springer-Verlag, Berlin, 1985.

Gipson, G. S., and Camp, C. V. "Phreatic Surface and Subsurface Flow With Boundary Elements Using an Advanced Green's Function." Betech 86: Proc. of the 2nd Boundary Element Technology Conference. Massachusetts Institute of Technology, June, 1986; Computational Mechanics Publications, Southampton, 1986.

Gipson, G. S. "Use of Residue Theorem in Locating Points Within an Arbitrary Multiply-Connected Region." Advances in Engineering Software, Vol. 8, No. 2, 1986, pp. 73-80.

Gipson, G. S., Reible, D. D., Savant, S. A. "Boundary Elements and Perturbation Theory for Certain Classes pf Hyperbolic and Parabolic Problems." Boundary Elements IX, Vol 3., 1987, pp. 115-127.

Gipson, G. S. Boundary Element Fundamentals - Basic Concepts and Recent Developments in the Poisson Equation, Topics in Engineering Vol. 2, Computational Mechanics Publications, Southampton, U.K., 1987.

Gradshteyn I. S., and Ryshik, I. M. Tables of Integrals, Series, and Products. Academic Press, New York, 1980.

Guo-Shu S. and Mukherjee S. "Boundary Element Method Analysis of Bending of Elastic Plates of Arbitrary Shape with General Boundary Conditions". Engineering Analysis, Vol. 3, NO. 1, 1986, pp. 36-44.

Hansen E. B. "Numerical Solution of Integro-Differential and Singular Integral Equations for Plate Bending Problems." Journal of Elasticity, Vol. 6, 1976, pp. 39-56.

Hildyard M. L., Ingham, D. B., Heggs, P. J., and Kelmanson, M. A. "Integral Equation Solution of Viscous Flow Through a Fibrous Filter " Boundary Elements VII: Proc. of the Seventh Intl. Conference on Boundary Element Methods. September, 1985, Lake Como, Italy; Springer-Verlag, Berlin, 1985.

Hoagland, D. A., and Prud'homme, R. K. "Taylor-Aris Dispersion Arising From Flow in a Sinusoidal Tube." AIChE Journal, Vol. 31, 1985., pp. 236-244.

Ingham, D. B., and Kelmanson, M. A. "A Boundary Integral Equation Method for the Study of Slow Flow Within Bearing Geometries." Proceedings of the 5th International Conference on Boundary Elements. Hiroshima, Japan; Springer-Verlag, Berlin, 1983.

Ingham D. B., and Kelmanson M. A. Boundary Integral Equation Analyses of Singular, Potential, and Biharmonic Problems, Lecture Notes in Engineering, Vol. 7, Springer-Verlag, Berlin, 1984.

Ingham D. B., Hildyard, M. L., Heggs, P. J. "Flow Through a Cascade." Boundary Elements IX, Vol. 3, 1987, pp. 443-457.

Jaswon, M. A., Maiti, M., and Symm, G. T. "Numerical Biharmonic Analysis and Some Applications." International Journal of Solids and Structures, Vol. 3, 1976, pp. 309-332.

Jaswon, M. A., and Maiti, M. "An Integral Equation Formulation of Plate Bending Problems." Journal of Engineering Mathematics, Vol. 2, 1968, pp. 83-93.

Jaswon, M. A., and Symm G. T. Integral Equation Methods in Potential Theory and Elastostatics. Academic Press, London, 1977.

Katsikadelis, J. T., and Armenakas, A. E. "Analysis of Clamped Plates on Elastic Foundation by the Boundary Integral Equation Method." Journal of Applied Mechanics, ASME, Vol. 51, 1984, pp. 574-580.

Katsikadelis, J. T., and Armenakas, A. E. "Numerical Evaluation of Double Integrals With a Logarithmic of Cauchy-Type Singularity." Journal of Applied Mechanics, ASME, Vol. 50, 1983, pp. 682-684.

Katsikadelis, J. T., and Armenakas, A. E. "Plates of Elastic Foundation by BIE Method." Journal of Engineering Mechanics, ASCE, Vol. 110, 1984, pp. 1086-1105.

Katsikadelis, J. T., and Armenakas, A. E. "Numerical Evaluation of Line Integrals With a Logarithmic Singularity." AIAA Journal, Vol. 23, 1984, pp. 1135-1137.

Katsikadelis, J. T., and Kallivokas, L. F. "Clamped Plates on Pasternak-Type Elastic Foundation by the Boundary Element Method." Journal of Applied Mechanics, ASME, Vol. 53, 1986, pp. 909-917.

Kellogg, O. D. Foundations of Potential Theory. Dover Publications, Inc., New York, 1954.

Kelmanson, M. A. "Boundary Integral Equation Solution of Viscous Flows With Free Surfaces." Journal of Engineering Mathematics, Vol. 17, 1983(a), pp. 329-342.

Kelmanson, M. A. "An Integral Equation Method for the Solution of Singular Slow Flow Problems." Journal of Computational Physics, Vol. 51, 1983(b), pp. 139-158.

Kerr, A. D. "Elastic and Viscoelastic Foundation Models." Journal of Applied Mechanics, ASME, Vol. 31, 1964, pp. 491-498.

Kreyszig, E. Advanced Engineering Mathematics. Fifth Ed. Wiley & Sons, New York, 1983.

Lamb, H. Hydrodynamics. Dover Publications, Inc., New York, 1945.

Lapidus, L., and Pinder, G. F. Numerical Solution of Partial Differential Equations in Science and Engineering. John Wiley & Sons, New York, 1982.

Lebedev, N. N., Skalskaya, I. P., and Ulflyans, Y. S. Worked Problems in Applied Mathematics. R. A. Silverman, trans. Dover Publications, New York, 1965.

Lebedev, N. N. Special Functions and Their Applications. R. A. Silverman, trans. and ed. Dover Publications, Inc., New York, 1972.

Leissa, A. W., Lo, C. C., and Niedenfuhr, F. S. "Uniformly Loaded Plates of Polygonal Shape." AIAA Journal, Vol. 3, 1965, pp. 566-567.

Ligget, J. A., and Salmon, J. R. "Cubic Spline Boundary Elements." International Journal of Numerical Methods in Engineering, Vol. 17, 1981, pp. 543-556.

Maiti, M., and Chakrabarty S. K. "Integral Equation Solutions for Simply Supported Polygonal Plates." International Journal of Engineering Science, Vol. 12, 1974, pp. 793-806.

Mills, R. D. "Computing Internal Viscous Flow Problems for the Circle by Integral Methods." Journal of Fluid Mechanics, Vol. 79, 1977, pp. 609-624.

Morjaria, M., and Mukherjee, S. "Inelastic Analysis of Transverse Deflection of Plates by the Boundary Element Method." Journal of Applied Mechanics, Vol. 47, 1980, pp. 291-296.

Mukherjee, S. "A Boundary Element Formulation for Planar Time-Dependent Inelastic Deformation of Plates With Cutouts." Int. Journal of Solids and Structures, Vol. 17, 1981, pp. 115-126.

Ng, S. S. F. "Influence of Elastic Support on the Behavior of Clamped Plates." Developments in Mechanics, Vol. 5, Proc. 11th Midwestern Mechanics Conference, 1969, pp. 343-371.

Ortiz, J. C. "An Improved Boundary Element Analysis System for the Solution of Poisson's Equation." M.S. thesis, Louisiana State University, 1986.

Ortiz, J. C., Walters, H. G., Gipson, G. S., and Brewer, J. A. III. "Development of Overhauser Splines as Boundary Elements." Boundary Elements IX, Vol. 1, 1987, pp. 401-407.

Overhauser, A. W. "Analytic Definition of Curves and Surfaces by Parabolic Blending." Ford Motor Company Technical Report, SL68-40, 1968.

Press, W. H., Flannery, B. P., Teukolsky, S. A., and Vetterling, W. T. Numerical Recipes. Cambridge University Press, Cambridge, 1986.

Rouse, H. Elementary Mechanics of Fluids. Dover Publications, Inc, New York, 1978.

Schlichting, H. Boundary-Layer Theory. J. Kestin, trans. McGraw-Hill, New York, 1979.

Segedin, C. M., and Brickell, D. G. A., "Integral Equation Method for a Corner Plate." Journal of the Structural Division, ASCE, Vol. 94, No. ST1, 1968.

Selvadurai, A. P. S. Elastic Analysis of Soil-Foundation Interaction. Elsevier /North-Holland, 1979.

Slattery, J. C. Momentum, Energy, and Mass Transfer in Continua. Robert E. Krieger Publishing Company, New York, 1981.

Stern, M. "A General Boundary Integral Formulation for the Numerical Solution of Plate Bending Problems." International Journal of Solids and Structures, Vol. 15, 1979, pp. 769-782.

Stern, M. "Boundary Integral Equations for Bending of Thin Plates." Progress in Boundary Element Methods. Vol. 2. C.A. Brebbia, ed. Pentech Press, London, 1983.

Stroud, A. H., and Secrest, D. Gaussian Quadrature Formulae. Prentice-Hall, Englewood Cliffs, New Jersey, 1966.

Syngellakis, S., and Kang, M. "A Boundary Element Solution of the Plate Buckling Problem." Engineering Analysis, Vol. 4, No. 2, 1987, pp. 75-81.

Szilard, R. Theory and Analysis of Plates--Classical and Numerical Methods. Prentice-Hall, New York, 1974.

Telles, J. C. F. The Boundary Element Method Applied to Inelastic Problems. Lecture Notes in Engineering. Vol. 1., Springer-Verlag, Berlin, 1984.

Timoshenko, S. T., and Woinowsky-Krieger, S. Theory of Plates and Shells. McGraw-Hill, New York, 1959.

Tottenham, H. "The Boundary Element Method for Plates and Shells." Developments in Boundary Element Methods 1. P. K. Banerjee and R. Butterfield, ed. Applied Science Publishers, Ltd., London, 1979.

Ugural, A. C. Stresses in Plates and Shells. McGraw-Hill, New York, 1981.

Walters, H. G. "Techniques for Boundary Element Analysis in Elastostatics Influenced by Geometric Modelling " M.S. thesis, Louisiana State University, 1986.

Weaver, W., and Johnson, P. R. Finite Elements for Structural Analysis. Prentice-Hall, Englewood Cliffs, pp. 18-21, 1984.

Wu, B. C., and Altiero, N. J. "A Boundary Integral Method Applied to Plates of Arbitrary Plan Form and Arbitrary Boundary Condi-tions." Computers and Structures, Vol. 10, 1979, pp. 703-707.

Wylie, C. R., and Barrett, L. C. Advanced Engineering Mathematics. McGraw-Hill, New York, 1982.

Xu, B., and Hansen, E. B. "Transient Stokes FLow in a Wedge." Journal of Applied Mechanics, Vol.54, 1987, pp. 203-208.

Youngren, G. K., and Acrivos, A. "Stokes Flow Past a Particle of Arbitrary Shape: A Numerical Method of Solution." Journal of Fluid Mechanics, Vol. 69, 1975, pp. 377-403.

Zienkiewicz, O. C. The Finite Element Method. 3rd Ed. McGraw-Hill, Maidenhead, U.K., 1977.

Lecture Notes in Engineering

Edited by C.A. Brebbia and S.A. Orszag

For information about Vols. 1-39 please contact your bookseller or Springer-Verlag.

Lecture Notes in Engineering

Edited by C. A. Brebbia and S. A. Orszag